Manual of Soil, Plant and Water Analysis

THE AUTHORS

Dr. Tahir Ali started his career in 1993, as Soil Surveyor in Water and Power Consultancy Services India Limited (WAPCOS), thereafter worked for Maharashtra Hybrid Seed Co. Ltd., India (MAHYCO) as Officer quality control from 1995 to 1998. He joined Sher-e-Kashmir University of Agricultural Sciences and Technology of Kashmir as Assistant Professor, in the Division of Soil Science on 24-11-1998 and is presently working as Associate Professor since 03-07-2002. He has 10 years of teaching experience and has guided 3 M.Sc. and 2 Ph.D students. He has published 25 Research papers in various journal of repute.

Shri Sumati Narayan has passed B.Sc. (Ag.) from Kanpur University and M.Sc. (Ag.) from Agronomy in Meerut University in 1988. He worked as an Extension Officer in a NGO for two years, about one year as SMS thereafter he joined as SRF in Agronomy Department of G.B.PUA&T Pantnagar in 1992. He came in Sher-e-Kashmir University of Agricultural Sciences and Technology of Kashmir as Assistant Professor (Vegetables) in April, 2002. He possess 8 years experience of extension, 16 years of research. He has published 20 research papers in various journal of repute.

Manual of Soil, Plant and Water Analysis

Dr. Tahir Ali
Associate Professor,
Division of Soil Science,
SKUAST-K, Shalimar

Mr. Sumati Narayan
Assistant Professor,
Division of Olericulture,
SKUAST-K, Shalimar

2016
Daya Publishing House®
A Division of
Astral International Pvt. Ltd.
New Delhi - 110 002

ISBN: 978-93-5124-098-3 (International Edition)

Published by   : **Daya Publishing House**®
     A Division of
     Astral International Pvt. Ltd.
     – ISO 9001:2008 Certified Company –
     4760-61/23, Ansari Road, Darya Ganj
     New Delhi-110 002
     Ph. 011-43549197, 23278134
     E-mail: info@astralint.com
     Website: www.astralint.com

Forword

The most valuable part of the mother earth is soil which as a resource base for life on the planet. Thus, we have to maintain this important natural resource for our own survival. For this, the deficits generated on various accounts have to be made up. This is only possible when we know about the real situation prevailing at a point of time. Soil analysis come into picture here. I am happy that Dr. Tahir Ali and Shri Sumati Narayan have endeavoured in this direction by authoring the book entitled *"Manual of Soil, Plant and Water Analysis"*. The book has been written in a lucid and self explanatory manner. I do hope that the book will be quite useful to all concern including students, teachers, researchers and quality control personals.

Prof. Anwar Alam,

Vice-Chancellor,

S.K. University of Agril. Sciences and Technology of Kashmir,

Shalimar–191121 (J&K)

Preface

Soil, plant and water are the major ingredients of Agriculture system. Analysis of soil, plant and irrigation water as natural resources is of utmost importance for conservation strategies. To make things easy for them, the authors have attempted to bring all pertinent information together in the present work. The principal involved in various determinations have been explained and substanciated with modern concepts, giving chemical equations where necessary. The methods are described under five subheads *viz.*, principles, apparatus, reagents, procedure, observations and calculation. The book contains simple and reliable procedure which can be useful to students, teachers, scientist and analysed in the field of applied sciences specially in the field of Soil Science, Agronomy, Horticulture and Environmental Science. Efforts are made to make the publication self explanatory and useful.

Finally, we would like to express our sincere gratitude to all the readers and also request them to send their valuable constructive remarks for the improvement in the next edition will be highly appreciated.

Tahir Ali

Sumati Narayan

Contents

Chapter 1
Soil Testing and Fertility Management

1.1 Definition

Soil is the storehouse for providing nutrients to the crops plants. The amount and kind of nutrients required for a particular crop vary from soil to soil and even from field to field on apparently the same type of soil. The extent to which a soil can supply nutrients to a crop grown decides the amount of fertilizers needed to compensate the deficiency. The nutrients requirement of a crop can only be estimated by soil testing. The use of fertilizers and amendments without testing the soil is like taking medicine without consulting physician. The soil testing thus may be defined as rapid, chemical and other test made on a soil for assessing the status of available nutrients and other edaphic properties having direct bearing on its management and to find out what and how much is needed.

1.2 Objectives

Main objectives of the soil testing are:

☆ To determine the available nutrients status of a soil and to test other common edaphic soil factors such as texture, soil reaction, salinity, calcareousness and organic matter status.

☆ To indicate the seriousness of any deficiency, toxicity or imbalance of nutrients in the soil.

☆ To form the basis on which fertilizer, manure and amendment needs are determined.

☆ To make fertilizer recommendations for most economic and optimum yield of crops.

☆ To serve as a guide for selection of fertilizers and best timings and methods of applications for a particular soil and crop to affect maximum use efficiency of soil and fertilizer nutrients.

☆ To help in selection of proper crops for a particular soil in extreme conditions.

☆ To group soils into various fertility classes and prepare fertility maps of a particular area or region.

1.3 Importance and Facilities

High crop yield can not be obtained without applying sufficient fertilizers to overcome existing deficiencies. The fertilizer being a costly input, each kilogram of it must be used very judiciously. While more and more fertilizer must be utilized if crop yields are to be increased, it is also essential that for each rupee the cultivator spends for plant food he receives, in return a substantial gain. Thus, soil testing programme plays an important role for farmers in adopting better and economic use of fertilizers and improved soil management practices for increasing agricultural production.

For efficient use of fertilizers it is necessary to know the right amount, right type, proper method of application, right timings of application and suitable combination of fertilizers with manures and organic wastes. Various soil tests serve as guide in all these aspects of fertilizers use. In addition, the amendment needs in respect of type and quantity of material to be applied in acidic, alkali and salt affected soils are also indicated by the soil testing. In certain conditions the need for selecting acid, salt and alkali tolerant crops is also indicated by soil testing.

Biological methods for evaluating soil fertility have certain advantages but most of these tests have serious disadvantage of being time consuming, hence the same are not widely adopted for use with large numbers of farmers. A chemical soil test, on the other hand is much more rapid and has the added advantage over method of deficiency symptoms and plant analysis in that soil test determines the needs of the soil or field before the crop is planted. In India, the soil testing service was started under the Indo-US operational agreement (1952) with the establishment of 24 soil testing laboratories in various parts the country. The I.A.R.I. carried out the basic research needed for selecting the most suitable soil test procedures for the use of the different laboratories located in different soil climatic zones. At present there are more than 300 soil testing laboratories and about 60 mobile soil testing vans which are run by agencies like Central Govt., State, Govt., Fertilizer firms, Agricultural Universities and Colleges. Most of these agencies test the soil sample free while others charge very nominal token fee per soil sample.

1.4 Methods of Soil Fertility Evaluation

The classification of various methods for soil fertility evaluation are given in Figure 1.1.

1.4.1 Microbial Test for Determining Soil Fertility

The growth of microorganism is very much related to the fertility of the soil. The growth of microorganism is an indication of better soil fertility. These experiments are conducted either in solution or in soil system. There are two basic factors while assessing soil fertility.

(*a*) Physiological factors

(*b*) Requirement of nutrient by microorganisms

(*a*) Studies of Physiological Factors

(*i*) *Ammonifying power*–the extent of ammonia (NH_4^+) production is the index of soil fertility status.

Figure 1.1: Methods of Soil Fertility Evaluations

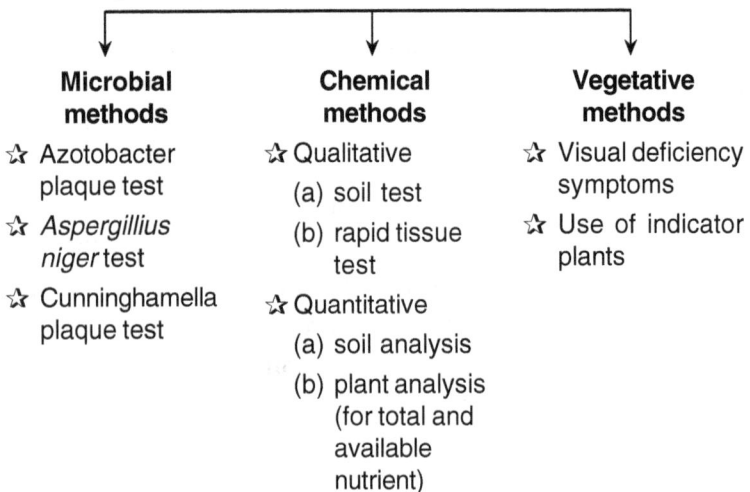

Microbial methods	Chemical methods	Vegetative methods
☆ Azotobacter plaque test	☆ Qualitative (a) soil test (b) rapid tissue test	☆ Visual deficiency symptoms
☆ *Aspergillius niger* test		☆ Use of indicator plants
☆ Cunninghamella plaque test	☆ Quantitative (a) soil analysis (b) plant analysis (for total and available nutrient)	

Weaker the power of decomposition, lesser is the ammonium production and hence poor soil fertility status.

(*ii*) *Cellulose decomposing power*: This is an index of soil fertility status because the amount of cellulose decomposed reflect the extent of available nitrogen and phosphorus in the soil.

(*b*) Requirement of Nutrient by Microorganisms

(*i*) *Azotobacter* plaque test

(*ii*) *Aspergillus niger* test

(*iii*) *Canninghamella* plaque test

1.4.1.1 *Azotobactor Plaque* Test

Azotobactor plaque test is conducted to find out lime requirement as they are very sensitive to acidity and to a low level of phosphorus. Advantage of this test is to determine the deficiencies of lime, phosphorus and potassium.

1.4.1.2 *Aspergillus niger* Test

The test is being used to determine potassium, copper, magnesium, molybdenum, cobalt and manganese requirements of the plants. The colour of the mycelia and spores is used as a measure of the amounts of copper or manganese present in the soil.

1.4.1.3 Cunninghamella Plaque Method

The method is used for determining phosphate content in the soils. The diameter of the conninghamella mycelia is used as criteria for the amount of phosphorus present in the soil.

1.4.2 Chemical Methods for Evaluating Soil Fertility

The chemical methods employed for determining soil fertility status include both qualitative and quantitative analysis of soil and plants.

1.4.2.1 Qualitative Test

1.4.2.2 Rapid Plant Tissue Tests

Rapid plant tissue tests are colorimetric diagnosis tests for estimating the soluble forms of major plant nutrients, *viz.*, N, P and K. In a situation where soil management systems are poor, yield and quality may gradually decline for many years before visual deficiency symptoms appear. Hence, plant tissue test can be used, as an aid in preventing the nutritional troubles.

Tissue testing can be particularly useful in detecting mild deficiencies before they become acute, and deficiency symptoms appear. The test can be used at any time during the growth season and thus provides a reliable basis for planning soil management and fertilizer recommendations.

Test for Nitrogen

Reagents

Diphenylamine Indicator

Dissolve 1 g of diphenylamine in 100 ml of concentrated H_2SO_4–when the solution becomes discoloured, prepare a fresh one.

Procedure

A. For Maize Crop

☆ Cut a thin vertical section at the plant node

☆ Add a drop of the colourless diphenylamine reagent.

☆ Study the colour after 30 seconds.

B. For *Wheat, Oat, Barley, Bajra, Soybean, etc.*

☆ Uproot the plant, and cut the stem near the lower node at an angle.

☆ Add 2-3 drops of diphenylamine reagent

☆ Note the intensity of the blue colour

No colour	Severally deficient
Slightly blue	Deficient
Moderately blue	Slightly deficient
Dark blue	Sufficient (No deficiency)

Test for Phosphate

Reagents

Reagent 1: Dissolve 8 g of ammonium molybdate in 200 ml of distilled water. To this add slowly with constant stirring 126 ml of conc. HCI and 874 ml of distilled water.

The concentrate reagent I should be diluted with 4 volumes of distilled water before use. The diluted reagent may become unsuitable for use after a few weeks, and should be checked by running a blank determination, before using it on plant material.

Reagent 2: Dry stannus chloride.

Procedure

☆ Cut leaf blades into fine pieces after removing the midrib

☆ Place a tea spoonful of the finely cut tissue in a flatbottomed vial, which has a 10 ml graduation.

☆ Fill the vial upto the 10 ml mark with phosphate reagent 1, and shake vigorously for one minute.

☆ Add a small amount of stannous chloride, mix the contents, and observe the colour.

No colour or yellow	Plants highly deficient in P (need to apply P-fertilizer)
Green or bluish green	Plant is deficient in P (Need to apply P-fertilizer)
Light blue	Plant has medium P (slight increase in yield expected with P-application)
Medium blue	Plant adequately supplied with P (no need to apply P-fertilizer)
Dark blue	Plant abundantly supplied with P (no need to supply P-fertilizer)

Test for Potassium

Reagents

Reagent 1: Dissolve 5 g of sodium cobaltinitrite and 30 g of sodium nitrite in 60 ml of distilled water; after the substances have dissolved, add 5 ml of glacial acetic acid, make up to 100 ml, and allow to stand for several days. Add 5 ml of this solution to a solution of 15 g of sodium nitrite in 100 ml of distilled water (15 per cent solution), and adjust to pH 5 with acetic acid.

Reagent 2: ethyl alcohol (95 per cent)

Procedure

☆ Cut leaf tissue into fine pieces

☆ Place 1/4th tea spoonful of the finely-cut leaf tissue in a glass vial, and 10 ml of potassium reagent 1.

☆ Shake vigorously for a minute carefully

☆ Add 5 ml of 95 per cent ethyl alcohol and mix

☆ After 2 to 3 minutes, observe the amount of yellow orange precipitate formed.

Only a trace of turbidity	Deficient K supply
Medium turbidity	Doubtful K supply
Very high turbidity	Adequate K supply

1.4.3 Quantitative Test

After taking composite/representative soil and plants, chemical analysis is taken up for a quantitative determination of N, P, K and other nutrients. The methods in detail are discussed in the Chapter 2 and 3 for soil and plant respectively.

1.4.4 Vegetative Methods

1.4.4.1 Visual Diagnosis of Deficiency Symptoms

This is based on the deficiency symptoms exhibited by crop plants (Figure 1.2). Visual diagnosis is not correct all the times because there are many symptoms that are common to more than one elements. Even then, one can obtain a hint about the deficiency of an element in plant as well as in the soil.

A. Effect Localized on Older or Lower Leaves

Nitrogen

☆ Sickly yellowish green colour

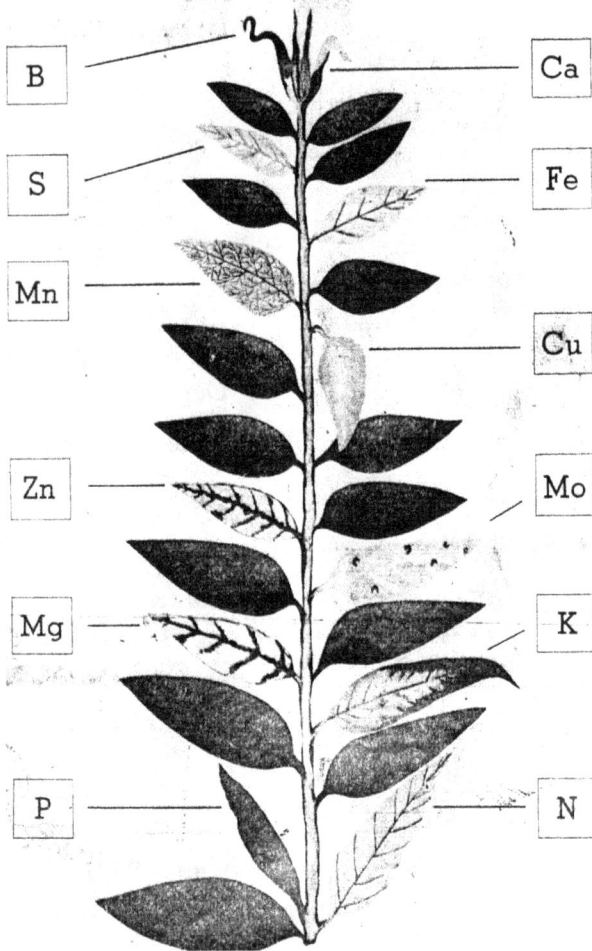

Figure 1.2: Nutrient Deficiency Symptoms in Plants

☆ Slow and dwarfed growth

☆ Drying up of leaves which starts at the bottom

Phosphorus

☆ Purplish leaves, stem and branches

    ☆ Slow growth

    ☆ Late maturity

Potassium

    ☆ Lower leaves scorched, burned on margin/tips, leaves curl and thicken

    ☆ Necrosis of leaves with the midrib remaining green

Magnesium

    ☆ Chlorosis

    ☆ Veins remain green

    ☆ Dead areas develop suddenly between the veins

Zinc

    ☆ Yellow striping of the leaves between the veins, older leaves die

B. Effect Localized on New Leaves

(a) Growing Tips Usually Dead

Calcium

    ☆ Wrinkled leaves in the terminal bud

    ☆ Death of roots, root's are short and branched

    ☆ Light green band along the margin of leaves

Boron

    ☆ Stems and leaf brittle

    ☆ Death of root tips

    ☆ Rossette appearance of leaves

(b) Growing Tips Remaining Alive

Iron

    ☆ Yellowing of leaves between veins with the veins remaining green

☆ Chlorosis, death of margins and tip of leaves

Manganese

☆ Pale leaves with scattered dead spots

☆ Veins remain green

Sulphur

☆ Leaves light green bed more than veins

☆ Drying of older leaves

☆ Immature fruit light green in colour

Copper

☆ Young leaves permanently wilted without spotting as marked chlorosis

Molybdenum

☆ Necrotic spots between the veins

1.4.4.2 Use of Indicator Plants

Indicator plants are very sensitive for particular nutrient and develop clear deficiency symptoms which are not shown by the crop. Some of the indicator plants are mentioned below.

Plants Elements	Indicator Plants
N and P	Cauliflower, cabbage
P	Rape seed mustard
K and Mg	Potato
Fe	Cauliflower, cabbage
Na	Sugar beet
Mn	Sugar beet, oat
B	Sunflower

1.5 Phases of Soil Testing

Time

One or two months before the growing of a crop is considered as the best. In standing crops soil sampling could be done between the rows after crop has attained physiological maturity.

Area

Upto two hectare sized plots with uniform management could be selected as a unit for collecting a representative soil sample. If the plot is of large size, it should be divided into small plots. In case of uneven topography, the size of the unit can be reduced as per the need.

Depth

The surface soil having a depth of 0-20 cms in mechanized farm and 0-15 cm in unmechanized farms is considered to be ideal for common field crops for sampling. In case of forest/orchard soil sample should be collected after digging a profile with a dimension of 90 x 120 x 150 cm and sample should be collected horizon-wise.

1.5.1 Instrument Used

Soil augers of different types *e.g.* screw auger, tube auger, post hole auger and khurpi or spade are used for taking the soil sample. In a very friable soil, a large spoon can be used (Figure 1.3).

1.5.2 Sampling Procedure

☆ Each field should be samples separately. When the ares within the field distinctly differ in crop growth, appearance of the soil, elevation or cropped differently, the field, should be divided and each area sampled separately (Figure 1.4).

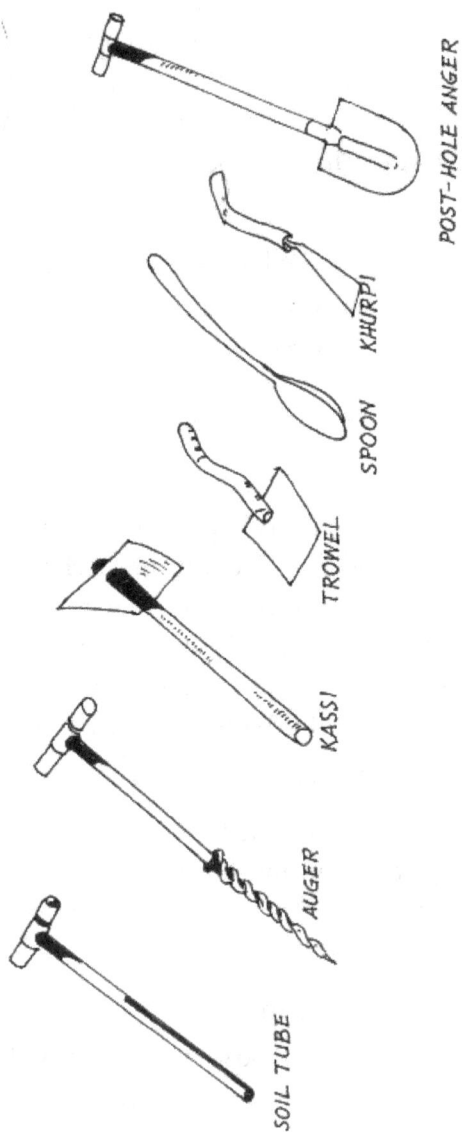

Figure 1.3: Sampling Tools for Soil Sample Collection

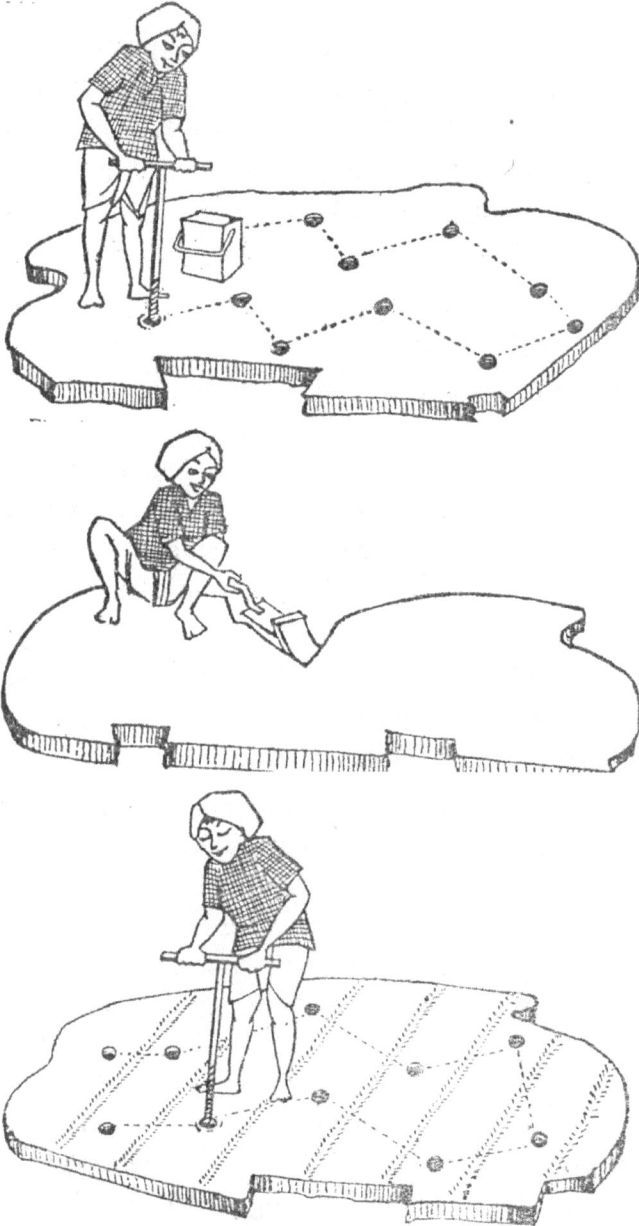

Figure 1.4: Collection of Soil Sample

☆ Drawing samples from the spots which do not represent the field should be avoided. Such spots may be bunds, marshy land, hedges, shady areas near irrigation channels or previously occupied by manure. Sampling should not done in a field within three months of the applications of lime or fertilizer.

☆ Proper sampling tools (Figure 1.3) should be used for the purpose.

☆ A composite sample may be taken from each area. After scrapping the surface litter, a uniform core or a thin slice of soil from the plough depth (0-20 cm) should be taken. V-shaped pit through plough depth is dug and slice from sides be taken (Figure 1.4). Augers are driven by rotary movement upto desired depth and the soil sample collected.

☆ Where crops have been planted in lines sampling may be done between the lines (Figure 1.4) at maturity stage of standing crop.

☆ Individual slice should be collected in a clean container (a tray, can etc.). The size of the sample should be reduced by successive quantity to about half a kilogram (Figures 1.5 and 1.6). Care should be taken to mix the composite soil thoroughly well each time during halving or quartering.

1.5.3 Dispatch

The representative soil sample is packed in a clean cloth bag. A tag indicating the address of the farmers, plot number and other details about cropping pattern etc. is attached with sample (preferably one keeping inside the dispatch-

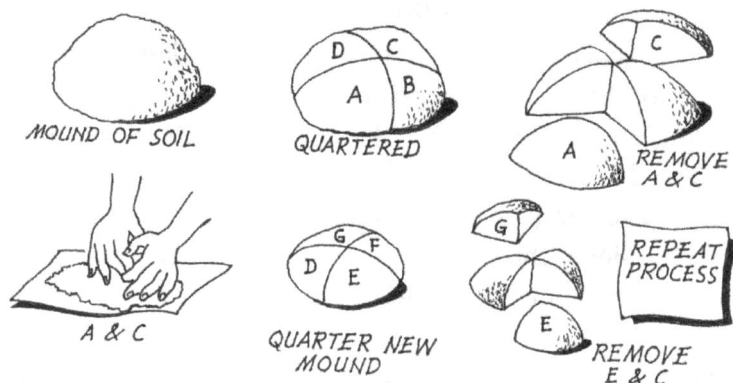

Figure 1.5: Reducing Procedure for Soil Sample

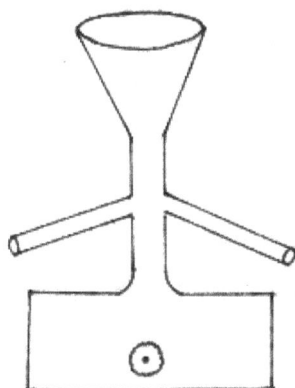

Figure 1.6: Soil Sample Reducer or Splitter

bag also). A performa giving the information about address of the farmer, plot number details of sampling and history of previous crop grown and crops to be grown including the desired yield from those crops is also filled up at the time of dispatch of the sample. The sample is dispatched to the near by soil testing laboratory for analysis.

1.5.4 Sample Preparation

After collection, the sample are normally left in a shady place to attain equilibrium with the moisture of the air. During air-drying, sample should be mixed (stirred with fingers) to expose fresh surfaces. After air drying soil samples are crushed gently in a pestle and mortar and sieved through a 2 mm sieve, the process being continued till the sieve contains no soil aggregates. Samples for micronutrient analysis should be crushed in a mortar of porcelain or stoneware and sieved through a stainless steel of nylon sieve. The air dry soil sample, passing through 2 mm sieve should be halved and quartered or processed repeatedly through a mechanical sample splitter (Figure 1.6) The sub-sample is the ground until it passes through a 0.5 mm sieve and transferred to a suitable container (tray or dish) for analysis.

1.5.5 Analysis

Generally soil is tested for the following physico-chemical properties:

1.5.5.1 Soil Texture

It guides on the following:

☆ Splits of fertilizer N doses in sandy soils should be increased.

☆ If soil texture is sandy having low pH use of Ammonium Chloride and Ammonium Sulphate be avoided.

☆ In very light or very heavy soils, use of FYM and compost in highly desirable.

☆ Sandy soils should be irrigated frequently but with small amount of water.

☆ In clayey soil phosphoric and potassium fertilizers should be placed and their broadcast should be avoided.

1.5.5.2 Electrical Conductivity

It guides on the following

☆ Index of salt content.

☆ Salinity hazards.

☆ Selection of salt tolerant crops.

☆ Need of amendments for the reclamation of saline and alkali soils

1.5.5.3 pH

It helps in:

☆ Selection of fertilizers and application methods.

☆ Selection of crops in extreme conditions.

☆ Selection of amendments.

☆ Diagnosis of deficiencies, toxicities and imbalance of nutrients.

1.5.5.4 Calcareousness

Help in:

☆ Choice of fertilizers.

☆ Diagnosis of nutrients deficiencies.

☆ Choice of Amendments.

1.5.5.5 Organic Carbon

It guides on the following:

☆ An index of nitrogen content.

☆ Structural status of soil.

☆ Water holding capacity.

☆ Buffering capacity.

☆ Need for FYM/Compost/ Green manuring.

☆ General fertility status of soil.

1.5.5.6 Available Nitrogen

1.5.5.7 Available Phosphorus

1.5.5.8 Available Potassium

1.5.5.9 Available Secondary Nutrients

1.5.5.10 Available Micronutrient (Zinc, Copper, Iron, Manganese, Boron and Molybdenum)

All of the above tests indicate the exent of availability of individual plant nutrient and guide to take decision about fertilizer practices.

1.5.5.11 Lime Requirement

Helps in deciding:

☆ The amount of lime to reclaim acidic soils.

1.5.5.12 Gypsum Requirement

Helps in deciding:

☆ The amount of gypsum required to reclaim saline and alkali soils.

1.6 Interpretation and Fertilizer Recommendations

From the results of analysis of soil samples soil test reports are prepared. Soil test reports are usually in three main parts. First part indicates the results of analysis of the soil samples. Second part is fertilizer recommendations for the crop based on soil test values. This part indicates quantities of nitrogen (N), phosphorus (P_2O_5), potash (K_2O), zinc and also lime or gypsum to be applied per hectare. The third part usually indicates time and methods of fertilizer application and other practices required to increase fertilizer use efficacy. Based on the results of these analysis, soil fertility maps may be prepared indicating the natural status of nitrogen, phosphorus, potash and zinc in the soils of an area. Once the soil from a field has been analyzed and a fertilizer recommendation is made, this recommendation can usually be adopted for three to four years without re-analysis the soil during this period. Recommendations include, as and when necessary, the quantity of lime or gypsum to be added in order to correct unfavourable soil conditions and in some cases also indicate the type of fertilizer to be used. These recommendations are not to be ignored for better soil management and higher productivity.

Chapter 2
Methods of Soil Analysis

2.1 Determination of Available Nutrients

2.1.1 Determination of Organic Carbon

Because of its complex nature, numerous difficulties beset the accurate estimation of soil organic matter. Organic matter is generally calculated from the determination of organic carbon on the assumption that organic matter, on an average in soil contains 58 per cent of carbon. Since, organic carbon can be determined directly with accuracy, it is preferably to report it as such, rather than a value of organic matter derived from it on the above assumptions. For routine work, rapid titration method is useful for indirect determination of nitrogen status of soil. However, the organic carbon status cannot be taken as a reliable index of available nitrogen, because the C : N ratio varies widely.

The organic carbon content of soils is estimated by using any of the following methods:

(a) Titrimetric determination (Walkley and Black, 1934).

(b) Colorimetric determination (Datta *et al.*, 1962). .

2.1.1.1 Titrimetric Method

This method oxidizes a lower percentage of the total organic carbon and only 77 per cent of the organic carbon get oxidized. Therefore, a correction factor should be considered in calculating the percentage of organic carbon.

Principle

The soil is treated with $K_2Cr_2O_7$ in presence of concentrate H_2SO_4 making use of the heat of dilution of sulphuric acid and the organic carbon in the soil is thus oxidizes to CO_2. The excess of chromic acid ($K_2Cr_2O_7$) not reduced by the organic matter, is determined by back titrated with standard ferrous ammonium sulphate solution using diphenyl amine indicator. In presence of phosphoric acid and diphenylamine indicator, the colour of solution changes from blue to green at the endpoint.

(A) The Oxidation of Carbon

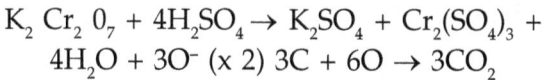

$$K_2 Cr_2 O_7 + 4H_2SO_4 \rightarrow K_2SO_4 + Cr_2(SO_4)_3 +$$
$$4H_2O + 3O^- \ (x \ 2) \ 3C + 6O \rightarrow 3CO_2$$

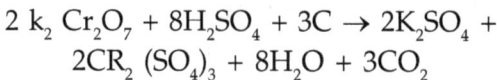

$$2 k_2 Cr_2O_7 + 8H_2SO_4 + 3C \rightarrow 2K_2SO_4 +$$
$$2CR_2 (SO_4)_3 + 8H_2O + 3CO_2$$

B) Titration

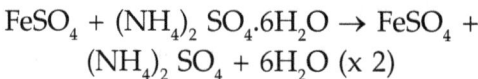

$$FeSO_4 + (NH_4)_2 SO_4.6H_2O \rightarrow FeSO_4 +$$
$$(NH_4)_2 SO_4 + 6H_2O \ (x \ 2)$$

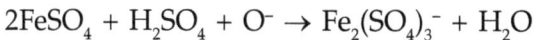

$$2FeSO_4 + H_2SO_4 + O^- \rightarrow Fe_2(SO_4)_3^- + H_2O$$

$$2FeSO_4 \ (NH_4) \ SO_4.12H_2O + H_2SO_4 + O- \rightarrow 2(NH_4)_2 \ SO_4 +$$
$$Fe_2 \ (SO_4)_3 + 13 \ H_2O$$

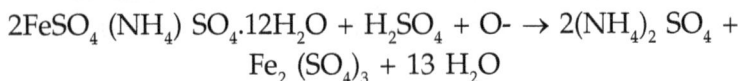

(C) Reaction of Diphylamine Indicator

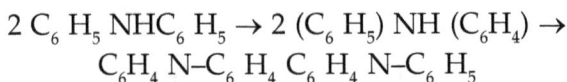

$$2 \ C_6 \ H_5 \ NHC_6 \ H_5 \rightarrow 2 \ (C_6 \ H_5) \ NH \ (C_6H_4) \rightarrow$$
$$C_6H_4 \ N-C_6 \ H_4 \ C_6 \ H_4 \ N-C_6 \ H_5$$

Apparatus

Conical flasks

Pipettes

Automatic pipette

Burette

Measuring cylinder

Reagents

Potassium Dichromate Solution (1 N): Dissolve 49.04gm of AR grade $K_2Cr_2O_7$ in 1 litre of distilled water

Concentrate Sulphuric acid

Orthophosphoric Acid (85 per cent)

Diphenylamine Indicator: Dissolve 0.5g diphenyl amine in 20 ml distilled water and add 100 ml conc. H_2SO_4

Ferrous Ammonium Sulphate (N/2): Dissolve 196 gm of $Fe(NH_3)_2 \ (SO_4)_2 \bullet 6 \ H_2O$ in 800 ml distilled water. Add 20 ml conc. H_2SO_4 and make the volume to 1 litre

Procedure

☆ Take 1.0 gm. Soil sample in 500 ml conical flask.

☆ Add 10 ml 1 N $K_2Cr_2O_7$ solution and 20 ml conc. H_2SO_4.

☆ Mix gently and allow the reaction to complete for 30 minutes.

☆ Dilute the reaction mixture with 200 ml distilled water and 10 ml H_3PO_4 (It checks the oxidation of diphenyl amine by Fe^{+3} ions after Fe^{+3} forming the ferric phosphate complex).

☆ Add 1 ml of diphenyl amine indicator for developing the colour.

A deep violet colour will appear. Titrate it against N/2 ferrous Ammonium sulphate solution, till the violet colour changes to green. Carry out a blank (without soil) in similar manner.

Observations and Calculation

☆ Weight of the soil = W g

☆ Volume of N/2 Fe $(NH_4)_2$ $(SO_4)_2$ needed for blank titration = X ml

☆ Volume of N/2 Fe $(NH_4)_2$ $(SO_4)_2$ needed for sample titration = Y ml

☆ Volume of 1 N $K_2Cr_2O_7$ used = 0.5 X (X-Y) ml

☆ Walkley averaged 77 per cent recovery of organic carbon (OC). Therefore, the correction factor is 100/77 = 1.3

☆ % organic carbon = $\dfrac{0.5 \times (X\text{-}Y) \times 1 \times 0.02 \times 100}{W}$

= A

Therefore, actual per cent of organic carbon = A x 1.3 = B

2.1.1.2 Colorimetric Method

Principal

The soil is digested with $K_2Cr_2O_7$ in presence of H_2SO_4

containing 1.25 per cent Ag_2SO_4. The remaining amount of K_2Cr_2O which is not utilized in the digestion of soil, produces chromium sulphate. The intensity of green colour is read on the photoelectric colorimeter after adjusting the blank solution to zero at 660 nm using red filter. Percentage of organic carbon is calculated from standard curve.

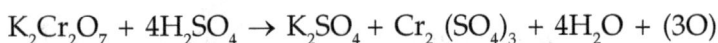

$$K_2Cr_2O_7 + 4H_2SO_4 \rightarrow K_2SO_4 + Cr_2 (SO_4)_3 + 4H_2O + (3O)$$

Apparatus

Colorimeter.

Conical flask.

Pipette

Measuring cylinder.

Centrifuge.

Reagents

1 N Potassium Dichromate Solution (49.04 g l^{-1})

Concentrate H_2SO_4 Containing 1.25 g Silver Sulphate (per 100 ml)

Procedure

☆ Take 1 g soil sample in 100 ml conical flask.

☆ Add 10 ml 1 N $K_2 Cr_2 O_7$ and stir a little then add 20 ml H_2SO_4 containing 1.25 per cent $Ag_2 SO_4$ with stirring.

☆ Keep for half an hour and centrifuge.

☆ Green chromium sulphate colour of the sample is read on the colorimeter.

Preparation of standard curve: Take 1 to 25 mg of anhydrous sucrose (AR quality) in different 100 ml conical flasks. Add 10 ml of 1 N $K_2Cr_2O_7$ in each flask, than add 20 ml conc. H_2SO_4, stir a little and keep for half an hour centrifuge and take the reading on colorimeter at 660 nm using red filter after adjusting to zero. Draw a curve between different concentrations of carbon and photoelectric colorimeter readings.

Calculation

% Organic Carbon = Colorimeter reading x 0.00421
(from standard curve)

2.1.2 Determination of Available Nitrogen

The method for available N determination in soil includes the following two forms of nitrogen:

1. Those easily available to the plants (minerals forms, NH_4^+ and NO_3)

2. Potentially available (minerals forms, NH_4^+, NO_3^- and organic nitrogen) some of the extractants are commonly used for estimation of these two forms of N are as follows:

   (*i*) Alkaline potassium permanganate method (Subbiah and Asija, 1956).

   (*ii*) Calcium hydroxide method

   (*iii*) Incubation method (Kenny and Bremmr, 1962).

   (*iv*) Nitrate nitrogen by phenoldisulphonic acid method

2.1.2.1 Alkaline Potassium Permanganate Method

Principle

Alkaline potassium per manganate extracts the inorganic nitrogen (NH_4^+ and NO_3^-) and readily oxidizable nitrogen from organic compounds. The extracted nitrogen is being distillated with sodium hydroxide, which librates NH_3 from the soil. The liberated NH_3 is being absorbed in boric acid containing mixed indicator. The amount of NH_3 absorbed is detrmined by titrating with standard HCl.

Apparatus

> Distillation flasks
>
> Conical flasks
>
> Balance
>
> Burette
>
> Measuring cylinder
>
> Distillation set

Reagents

Potassium Permanganate Solution (0.32 per cent): Dissolve 3.2 g of $KMNO_4$ in one litre of distilled water

Sodium Hydroxide Solution (2.5 per cent): Dissolve 25 g of NaOH pellets in one litre of distilled water

Boric Acid (4 per cent): Dissolve 4 g of boric acid in 900 ml of distilled water. Add 5 ml of mixed indicator and adjust the pH to 4.5 and finally make the volume to one litre with distilled water

Sulphuric Acid (N/50)

Paraffin Liquid

Mixed Indicator: Dissolved 0.5 g Bromo cresol green and 0.1 g of methyl red in 100 ml of 95 per cent ethnol

Procedure

> ☆ Take 10 gm of processed soil sample in a distillation flask and add 20 ml distilled water.

☆ Add 100 ml of 0.32 per cent $KMnO_4$ solution

☆ Add 1 ml paraffin liquid (anti-bumping agent).

☆ Before pouring NaOH take 50 ml 4 per cent boric acid in receiving flask and dip the delivery tube of distillation set in it. Now pour 100 ml 2.5 per cent NaOH in the distillation flask and fit it in the set. Continue distillation the distillate is free from NH_3.

☆ Determine the amount of NH_3 absorbed in boric acid, titrate it with $N/50$ H_2SO_4.

☆ Same way also carry out a blank.

Observations and Calculation

☆ Weight of soil = 10 g

☆ Volume of $N/50$ H_2SO_4 used for sample = S

☆ Volume of $N/50$ H_2SO_4 used for blank = B

$$N \ (kg/ha) = \frac{(S-B) \times N \times 2000}{W}$$

S = Volume of H_2SO_4 used for sample

B = Volume of H_2SO_4 used for blank

N = Normality of H_2SO_4

W = Weight of soil samples (g)

1000 ml of 1 N H_2SO_4 = 14 g N

1000 ml of $N/50$ H_2SO_4 = 14 × 0.2 g N

1 ml of $N/50$ H_2SO_4 = 14 × 2/1000 = (0.00028 g N)

2.1.2.2 Calcium Hydroxide Method

Principle

Potentially available nitrogen in soil is estimated by calcium hydroxide hydrolysis and thus NH_3 evolved is absorbed in 4 per cent boric acid. The amount of NH_3 is calculated by titrating with standard H_2SO_4.

Apparatus

　　Kjeldahl flasks

　　Conical flasks

　　Burette

　　Balance

　　Beakers

　　Measuring cylinders.

Reagents

　　Calcium Hydroxide [Ca $(OH)_2$]

　　Boric Acid (4 per cent) with mixed indicator (5 ml/liter of acid) adjusted to pH 4.5

　　Paraffin Liquid

　　Sulphuric Acid (N/50)

Procedure

　　☆ Take 20 g in a 800 ml Kjeldahl flask

　　☆ Add 1/4 teaspoonful of Ca $(OH)_2$. Dilute it with 200 ml distilled water.

　　☆ Add 2 ml of paraffin liquid.

　　☆ Take 25 ml of 4 per cent boric acid in receiving flask and dip the delivery tube in it. Fit the Kjeldahl flask in the distillation set and continue distillation till the distillate is free from NH_3.

　　☆ Titrate with standard H_2SO_4

　　☆ Simultaneously, run a blank sample (without soil).

Calculation

$$\text{Potentially available N (Kg/ha)} = \frac{(S-B) \times 0.014 \times N \times 2000000}{W}$$

where,

S = Volume of H_2SO_4 used for sample

B = Volume of H_2SO_4 used for blank

N = Normality of H_2SO_4

W = Weight of the soil.

2.1.2.3 Incubation Method (Kenny and Bremrer, 1962)

Principle

Soil is incubated for 14 days at 30°C after mixing the soil with acid washed quartz sand. After incubation 2 N KCl is mixed with soil and the amount of nitrogen as (NH_4 + NO_3 + NO_2) in the incubated sample is determined by steam distillation in the presence of MgO and deverdas alloy.

The amount of nitrogen (NH_4 + NO_3 + NO_2) in the sand soil mixture before incubation is also determined by the same procedure. Mineralizable N in soil sample is calculated by the difference of these two values.

Apparatus

Wide mouth bottle.

Incubator

Distillation unit

Measuring cylinders

Pipette

Burette

Reagents

  Potassium Chloride (2 N) solution

  Magnesium Oxide (MgO)

  Deverdas Alloy

  Boric Acid (2 per cent)

  Mixed Indicator: Dissolve 0.099 g. Bromocresol green and 0.066 g. Methyl red in 100 ml ethanol

Procedure

  ☆ Mix 10 g soil with 30 g of acid washed quartz sand and transfer to wide mouth bottle.

  ☆ Add 6 ml of distilled water and shake the bottle so that the mixture is distributed evenly.

  ☆ Close the bottle with a rubber stopper having a hole in the centre and place the bottle in the incubator for 14 days at 30°C.

  ☆ Add 100 ml 2 N KCl after incubation and shake for 30 minutes and then allow to stand.

  ☆ Pipette a clear 20 ml aliquot of the supernatant liquid in 100 ml distillation flask and determine the amount of N (NH_4 + NO_3 + NO_2) by steam distillation with MgO and deverdas alloy.

  ☆ Determine the N (NH_4 + NO_3 + NO_2) in the sand–soil mixture before incubation by the same process.

  ☆ Calculate the mineralizable N in soil sample from the difference between these two values.

2.1.2.4 Nitrate–N by Phenol Disulphonic Acid Method

Principle

  Nitrate in soil is extracted with 0.01 molar (M) $CuSO_4$ solution. Chloride interference is prevented by Ag_2SO_4.

Nitrate reacts with H_2SO_4 and produces yellow colour complex. The intensity of yellow colour is red at 415 nm using blue filter by spectrophotometer.

Reagents

Copper Sulphate Solution (0.5 M): Dissolve 15 g of $CuSO_4$ 5 H_2O in one litre of distilled water

Silver Sulphate Solution (0.6 per cent): Dissolve 6 g of Ag_2SO_4 in one litre of distilled water

Phenol Disulphonic Acid: Dissolve 25 g phenol +150 ml conc. H_2SO_4 (AR–NO_3 free) + 75 ml fumings H_2SO_4. Mix and heat on boiling water bath for 2 hours and store in a coloured bottle. If fuming H_2SO_4 is not available add same amount of conc. H_2SO_4 and heat on a water bath for 6 hours

Standard Nitrate Solution (100 ppm): Dissolve 0.7 22 g of KNO_3 in one litre of distilled water

Working standard nitrate solution (10 ppm): Dilute 100 ml of 100 ppm nitrate stock solution to one litre with distilled water

Apparatus

Colorimeter

Volumetric flasks

Measuring cylinder

Beakers.

Procedure

☆ Take 20 g of processed soil sample in a 500 ml conical flask.

☆ Add 50 ml distilled water and shake for one hour Add a pinch of $CaSO_4$ and shake for a minute and filter.

☆ Take 20 ml of aliquot in a 50 ml beaker or in an evaporating dish. Evaporate to dryness on water bath and cool.

☆ Add 3 ml of the phenol disulphonic acid and allow to react by rotating the container.

☆ After 10 min add 15 ml distilled water and stir with a glass rod.

☆ Allow it to cool and after it is cooled add 6 N NH_4OH slowly until the development of yellow colour.

☆ Add 3 ml NH_4OH and make the volume to 100 ml with distilled water.

☆ Read the intensity of yellow colour on the photoelectric colorimeter at 420 nm using blue filter.

Calculation

$$NO_3 \ N \ (mg/100g \ soil) = ppm \ in \ test \ solution$$
$$\times \ 100/20 \times 50/20 \times 1/100$$

2.1.2.5 Ammonium-Nitrogen by Colorimetric Method

Nitrogen, in the form of exchangeable ammonium accounts for 0.01 to 0.1 ml/100 g of soil, which is much lesser than the amount of N, determined as ammonium ion in soil. The ammonium ions, emanating from the soluble form, are subjected to either colorimetric determination or volumetric analysis, involving distillation with MgO. The colorimeter procedure, invariably, uses the Nessler reagent-potassium tetramercuriate ($1_4 \ HgK_2$), and a yellow complex of oxidimercurammonium iodine, $HgOHg \ (NH_2)I$ is formed. The intensity of the coloured complex is measured on a colorimeter at 410 nm.

Apparatus

Colorimeter

Micropore

Glass filter

Erlenmeyer flask

Buchner funnel

Reagent

Potassium Sulphate (0.2 N): Dissolve 2.239 of potassium sulphate in one of distilled water

Wash the solution in a 1 litre volumetric flask. Then, add 112 gm of KOH and bring the volume to about 800 ml. Mix cool and dilute 1 litre with distilled water. Allow the solution to stand for a few days, and decant off the clear supernatant liquid (Nessler's reagent) into amber coloured bottle for use

Sodium Tartarate (10 per cent, $Na_2C_4H_4O_5 \cdot 2H_2O$) solution

Preparation of Standard Curve: Dissolve 0.153 g of NH_4Cl in a few ml of distilled water in a 2 litre measuring flask, and the make up the volume to the mark with distilled water. This is the stock solution of 20 ppm

Take 0, 1, 2, 3, 4, 5, 7, 5 and 10 ml of the stock solution into 100 ml of volumetric flasks and add 12 ml of Na tartarate solution. Add water to make up about 93 ml total volume. Then, add 5 ml of Nessler's reagent with rapid mixing. Make up the volume, mix and read on a colorimeter

Procedure

☆ Take 20 g of soil in an erleymeyer flask and add 50 ml of 0.2 N K_2SO_4 solution.

☆ Agitate it for 10 minutes, through a micropore glass filter attached to a buchner funnel.

☆ Withdraw 20 ml of the extract and place it in a 100 ml volumetric flask.

☆ Add 2 ml of sodium tartarate solution

☆ Dilute the solution with distilled water, followed by addition of 5 ml of Nessler's reagent; homogenize the solution, and finally make up the volume to 100 ml with distilled water.

☆ Allow the solution to stand for 25 minutes and read the colour intensity at 410 nm by colorimeter

Observations and Calculation

☆ Weight of the soil = 20 gm

☆ Volume of K_2SO_4 taken = 50 ml

☆ Volume of aliquot used = 20 ml

☆ Total volume of extract = 100 ml

☆ Concentration of NH_4–N = Y ppm

NH_4N (per cent) in the soil sample = Y x 50/20 x 1/20 x 1/100

NH_4-N (ppm) in the soil sample = Y x 50/20 x 1/20 x 100

NH_4-N (kg/ha) in the soil sample = Y x 28

2.1.3 Determination of Available Phosphorus

Phosphorus in soil ranges from 0.01 to 0.3 per cent and present in several forms and combinations. The apitite group of primary minerals is the main source of soil phosphorus. The total amount of phosphorus present in soil is not available to the plants and only a fraction of it may be available to growing plant. For determining plant available phosphorus in soil two methods are commonly used. The Olsen's method (Olsen *et al.*, 1954) is used for neutral and alkaline soils while the Bray and Kurtz method (Bray and Kurts, 1945) is used for acid soils.

2.1.3.1 Olson's Method

Principle

Available P is extracted from the soil with 0.5 M $NaHCO_3$ (pH 8.5) and decolourising carbon is also added before shaking. In the filtrate ammonium molybdate is added. The yellow colour of ammonium complex is reduced to blue colour after adding stannous chloride. The intensity of blue colour is measured on calorimeter at 660 nm using red filter.

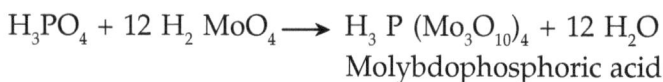

$$H_3PO_4 + 12\ H_2\ MoO_4 \longrightarrow H_3\ P\ (Mo_3O_{10})_4 + 12\ H_2O$$
$$\text{Molybdophosphoric acid}$$

Apparatus

Photoelectric calorimeter

Mechanical shaker

Conical flasks

Pipette

Funnels

Volumetric flasks.

Reagents

Sodium Bicarbonate Solution (0.5 M): Dissolve 42.0 g of sodium bicarbonate, $(NaHCO_3)$ in about 950 ml of distilled water, adjust the pH 8.5 using dilute HCl and make the volume to one litre with distilled water.

Activated Charcoal: (free from soluble phosphorus)

Ammonium Molybdate Solution: Dissolve 15 g ammonium molybdate in 300 ml of distilled water. Add 348 ml conc. HCl and dilutre to one litre with distilled water

Stannous Chloride: Dissolve 10 g of Sn Cl $_2$ in 25 ml conc. HCl and dilute 1 ml of this solution to 66 ml with distilled water.

Standard P Solution: Dissolve 0.1916 gm of Potassium dihydrogen orthophosphate ($KH_2 PO_4$) in 1 litre of distilled water. This solution contains 0.10 mg P_2O_5/ml. Take 10 ml of stock solution and dilute to 1000 ml with distilled water. This solution contains 0.001 mg/ml. Take 1, 2, 4, 6 and 10 ml working solution in 25 ml volumetric flasks. Add 5 ml of ammonium molybdate reagent, and dilute with distilled water to about 20 ml. Add 1 ml dilute $SnCl_2$ solution. Shake and make up the volume to 25 ml with distilled water. Read intensity of the colour on colorimeter at 660 nm using red filter. Plot the readings against concentration of P_2O_5.

Procedure

☆ Take 1 g of processed soil sample in 150 ml conical flask.

☆ Add 20 ml of 0.5 M sodium bicarbonate extractant and 1 g of activated charcoal.

☆ Shake for 30 minutes and filter by using Whatmans No. 42

☆ Take 10 ml of the filtered in a 50 ml volumetric flask

☆ Add 5 ml of the Ammonium molybdate solution. Dilute to about 20 ml with distilled water and add 1 ml of dilute Sn Cl$_2$ solution.

☆ Make up the volume with distilled water to 50 ml.

☆ Read the intensity of the blue colour on the colorimeter at 660 nm using red filter.

Observations and Calculation

☆ Weight of the soil = 1 g

☆ Volume of 0.5 M NaH CO_3 used = 20 ml

☆ 1st dilution = 20/1

☆ Volume of extractant taken = 10 ml

☆ Final volume made = 50

☆ Second dilution = 50/10

☆ Total dilution 20 x 5 = 100

☆ Transmittance of the test solution = T

☆ Concentration of the unknown solution as read from the standard curve = A

☆ Available P (ppm) = A x 100

☆ Available P (kg ha^{-1}) = Ax 100 x 2.24

Notes

☆ To convert lbs P_2O_5/acre to lbs P/acre, multiply by 0.44.

☆ To convert lbs/acre to kg/ha multiply by 1.12.

2.1.3.2 Bray's and Kurtz Method

This method is mostly used for acid soil (pH <5.5)

Principle

The soil is extracted with ammonium fluoride (0.03 N NH_4F + 0.025 N HCl) and the filtrate is treated with ammonium molybdate and the colour is developed by stannous chloride. The intensity of blue colour is measured on the photoelectric colorimeter at 660 nm using red filter.

Apparatus

Photoelectric colorimeter.

Mechanical shaker

Pipette

Funnel

Volumetric flasks

Reagents

Ammonium Fluride (0.03 N NH_4F + 0.025 NHCl): Dissolve 2.22 g. of Ammonium Fluoride in 200 ml of distilled water. Add 1800 ml water containing 4 ml of distilled conc. HCl, make up the volume to 2 lts with distilled water.

Ammonium Molybdate: Same as in Olsen's method.

$SnCl_2$: Same as in Olsen's method.

Standard P Solution: Same as in Olsen's method.

Procedure

☆ Take 5 g of processed soil sample in a 100 ml conical flask.

☆ Add 50 ml of the extractant.

☆ Shake for 5 minutes and filter.

☆ Take 5 ml of the filterate in 25 ml volumetric flask.

☆ Deliver 5 ml of the ammonium molybdate reagent and dilute to 20 ml with distilled water.

☆ Shake and add 1 ml of the dilute $SnCl_2$ solution.

☆ Make up the volume to 25 ml with distilled water.

☆ Read the intensity of blue colour at 660 nm using red filter.

☆ After calibrating also prepare the blank (without soil) in similar manner.

Observations and Calculation

☆ Weight of the soil = 1 g

☆ Volume of 0.5 M $NaHCO_3$ used = 20 ml

☆ 1st dilution = 20/1

☆ Volume of extractant taken = 10 ml

☆ Final volume made = 50

☆ Second dilution = 50/10

☆ Total dilution 20 x 5 = 100

☆ Tranmittance of the test solution = T

☆ Concentration of the unknown solution as read from the standard curve = A

☆ Available P (ppm) = A x 100

☆ Available P (kg ha^{-1}) = Ax 100 x 2.24

Quite a number of methods have been developed to obtain a readily soluble forms of inorganic phosphorus to represent the available soil P. The amount of available P content in soil is readily determined by using suitable reagents according to the specified soil to solution ratio and the shaking time. Some of these methods used for estimation of available P in soil are listed in Table 2.1.

2.1.4 Determination of Available Potassium

Potassium in the soil generally ranges from 0.05-3.5 per cent, out of which 95 per cent is in the complex form, 1-10 per cent relatively non-available and 2 per cent is available

Table 2.1

Sl.No.	Methods	Extract Composition	Soil : Extractant Ratio	Shaking Time (min)
1.	Morgan's method	10 ml of acetic acid (glacial) dissolved in 1 litre of 0.5 per cent NaOH (pH 8.4)	1: 5	30
2.	Mehlich's method (0.05 N HCl + 0.025 N H_2SO_4 (pH 1.2)	Prepared by mixing of 4.25 ml conc. HCl + 0.7 ml conc. H_2SO_4 in one litre of distilled water	1: 4	5
3.	Bray's I method (0.03 NH4 F + 0.025 N HCl (pH 3.5)	Dissolve 1.11g NH_4F in 2.1 ml of conc. HCl in one liter of distilled water	1: 10	5
4.	Bray's II method (0.03 N NH_4F + 0.1 HCl (pH 1.0)	Dissolve 1.11g NH_4F and 8.5 ml conc. HCl in 1 litre of distilled water	1: 20	2/3
5.	Olsen's method (0.5 M $NaHCO_3$ (pH 8.5)	Dissolved 42 g of $NaHCO_3$ in 1 litre of distilled water	1: 20	30

to the plants. The available K is usually determined by using neutral normal ammonium acetate solution in the soil.

Principle

Available K is extracted from the soil by shaking with 1 N Ammonium acetate solution. During the equilibrium, ammonium ions exchange with the exchangeable K ions of the soil. The K content in the equilibrium solution is estimated by Flame photometer.

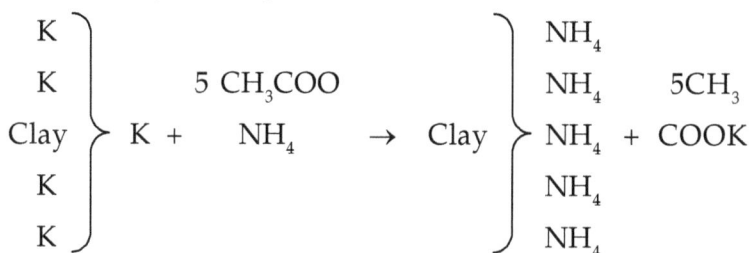

$$
\left.\begin{array}{c} K \\ K \\ Clay \\ K \\ K \end{array}\right\} K + \begin{array}{c} 5\ CH_3COO \\ NH_4 \end{array} \rightarrow Clay \left.\begin{array}{c} NH_4 \\ NH_4 \\ NH_4 \\ NH_4 \\ NH_4 \end{array}\right\} + \begin{array}{c} 5CH_3 \\ COOK \end{array}
$$

Apparatus

Flame photometer

Filter funnels.

Pipette

Mechanical shaker.

Reagents

Normal Neutral Ammonium Acetate Solution: Dissolve 77.09 g of Ammonium acetate (NH_4OAC) in distilled water and makeup the volume to one liter with distilled water (pH 7.0)

Standard K Solution: Dissolve 1.908 AR grade KCl in 1 litre of distilled water. It gives 1000 ppm K solution

Working K Standard Solution: Take 0, 5, 10, 15, 20, 30, 40 and 60 ml of this stock solution (1000 ppm K) separately and dilute to one litre with distilled water. These solutions contain 0, 5, 10, 20, 30, 40 and 60 ppm K, respectively

Procedure

☆ Take 5 g of soil in 100 ml conicle flask

☆ Add 25 ml of the neutral normal ammonium acetate solution

☆ Shake for 5 minutes on mechanical shaker and filter.

☆ Determines the K content on the flame photometer.

Observations and Calculation

☆ Weight of the soil = 5 g

☆ Volume of neutral normal ammonium acetate used = 25 ml

☆ Reading of the flame photometer for the test solution = A

☆ Concentration (ppm) as read from the standard curve = B

☆ Dilution factor = 25/5 = 5

☆ Available K in soil (ppm) = B x 5

☆ Available K in soil (kg ha^{-1}) = B x 5 x 2.24

☆ Available K_2O in soil (kg ha^{-1}) = B x 5 x 2.5 x 1.20

Note

1. To convert kg K_2O/ ha to kg K/ha. multiply by 0.83

2.1.5 Determination of Available Sulphur

Sulphur (S) occurs in numerous forms in soil, *viz.*, sulphites, sulphates, sulphides and in organic compounds.

The inorganic forms of sulphur is accessible to the plant. However, it is considered that the most accessible form is 'sulphate' (SO_4^{2-}). The organic forms of S compounds become assimilable, especially following microbiological transformation into SO_4^{2-}. The mineralization of organic S in a soil depends primarily upon the N: S ratio, and SO_4^{2-} may be fixed if Fe or Ba is present or if the soil is very acidic. Despite the fact that plants absorb S as SO_4^{2-}, mobility of SO_4^{2-} in soil may not always yield satisfactorily while assessing SO_4^{2-} availability. Development of chemical tests for the estimation of available S (SO_4^{2-}) is of recent interest in soil testing work, particularly in some reported S deficient areas, and for certain crops, whose requirement of this nutrients is high enough, often exceeding to that of P.

A number of extractants have been used for determination of available sulphur *viz.*, Olsen's, Morgan's, Bray's, normal magnesium acetate, and 0.15 per cent calcium chloride solution. Two procedures of extraction of S (SO_4^{2-}) are discussed here.

2.1.5.1 Monocalcium Phosphate Extractable S (Ensminger, 1954)

Principle

Soil is shaken with monocalcium phosphate solution, containing 500 ppm of P. During the extraction, phosphate ions displace the adsorbed sulphate ions. Calcium ions depress the extraction of soil organic matter, thus eliminating contamination from extractable organic S. This method extracts soluble SO_4^{2-}, plus a fraction of the adsorbed SO_4^{2-}. The filtrate is treated with barium chloride in the presence of gum acacia solution, and the turbidity of barium sulphate is measured by the turbidimetric method.

Turbidity Procedure

The filtrate is treated with barium chloride in the presence of gum-acacia and the turbidity of SO_4^{-2} as barium sulphate is measured colorimetrically gum acacia helps to prevent rapid settling of barium sulphate precipitate.

Apparatus

Photometric colorimeter,

Suction pump

Erylenmeyer flask

Pipette

Whatman No. 42 filter paper

Reagents

Extracting Solution: Dissolve 2.18 g of monocalcium phosphate (Ca $H_2PO_4)_2$ 2HO) in distilled water and dilute to 100 ml.

Barium Chloride: Grind $BaCl_2$ crystals in a mortar, until they pass through a 30 mesh sieve and are retained on a 60 mesh sieve. The crystals should be added to the sulphate solution. The size of the crystals will determine their rate of solution which, in turns, determines the rate of reaction with sulphate.

Standard Sulphur Solution: Dissolve 0.5434 g of the AR grade potassium sulphate in distilled water, and dilute to one litre. This gives 100 ppm S.

Preparation of the Standard Curve: Pipette out 0.25, 0.5, 1.0, 2.5, 5.0 and 7.5 ml of the stock solution in a series of 25 ml volumetric flasks. Add 10 ml of the extracting solution, in each flask followed by about 1 g of $BaCl_2$ crystals and shake for 1 minute. Then add 1 ml of 0.25 per cent of gum-acacia. Make up the volume with distilled water, and shake for 1 minute. These are the working S standards, with concentrations of 1, 2, 4, 10, 20 30 ppm S, respectively. Make turbidity measurements within a period, from 5-30 minutes following the formation of precipitate by a colorimeter at 420 nm using blue filter. Plot a curve between the concentration of S and the turbidity readings (per cent T or A).

Procedure

☆ Weigh 20 g of the air-dried soil, in a 250 ml erlenmeyer flask.

☆ Add 100 ml of the extracting solution, and shake for 30 minutes. Filter the contents through a Whatman No. 42 filter paper under suction.

☆ Transfer 20 ml of the aliquot to a 25 ml volumetric flask and then proceed further as described in the preparation of the standard curve.

Observations and Calculation

☆ Weight of the soil = 20 g

☆ Volume of the extractant added = 100 ml

☆ Volume of the aliquot taken = 20 ml

☆ Final volume = 25 ml

☆ Transmittance (per cent) as read from the colorimeter = T (say)

☆ S (ppm) from the standard curve = C (say)

☆ First dilution = 100/20 = 5 times

☆ Second edition = 25/20 = 1.25 times

☆ Total dilution = 5 x 1.25 = 6.25 times

Now, available S (ppm) in the soil   = C x 6.25

Available S (kg/ha) in the soil = C x 6.25 x 2.24

2.1.5.2 Turbidimetric Method (Massoumi and Cornfield, 1963)

A suitable aliquot is treated with concentrated barium sulphate seed suspension, containing barium chloride crystals in acidified medium, and the turbidity developed, by precipitating barium as barium sulphate, estimated by colorimeter at 440 nm using blue filter.

Apparatus

The same as used in the turbidimetric method

Reagents

Nitric Acid (25 per cent)

Acetic-phosphoric Acid: 900 ml of glacial (AR) acetic acid mixed with 300 ml of H_3PO_4

Gum Acacia-acetic Acid Solution: Dissolve 5 g of pure gum acacia in 500 ml of hot water and filter the hot solution through Whatman No. 42 filter paper. Cool the filtrate, and then dilute to 1 litre with dilute acetic acid.

Barium Sulphate Seed Suspension: Dissolve 18 g barium chloride (AR) in 44 ml of hot water. Add 0.5 ml of concentrated standard solution (given below). Bring to boiling and cool quickly. Put in it 4 ml of gum acacia-acetic acid solution. Prepare the seed suspension fresh every day before use.

Standard Sulphate Solution (2 mg S ml^{-1}): dissolve 1.09 g of oven-dried potassium sulphate (K_2SO_4) AR grade in distilled water, and dilute to 100 ml.

Working Sulphate Solution (10 µg S ml⁻¹): Dilute 5 ml of the standard sulphates solution (2 ppm) to 1 litre with distilled water.

Procedure

☆ Take 15 of the extract in a 25 ml volumetric flask.

☆ Add 2.5 ml of 25 per cent HNO_3 and 2 ml of acetic phosphoric acid and dilute to about 22 ml and shake.

☆ Add 0.5 ml of barium sulphate seed suspension (which must be shaken before use) and 2.0 g of barium chloride crystals, successively.

☆ Stopper the flask and invert 3 times. Invert 10 times after 10 minutes and 5 times after 5 minutes.

☆ Allow another 5 minutes thereafter add, 1 ml of gum acacia-acetic acid solution and dilute to volume (25 ml). Invert 3 times and set aside for one and half hour.

☆ After abovementioned 90 minutes invert the flask 10 times again, and read the turbidity on a colorimeter at 440 nm using blue filter.

Preparation of Standard Curve

Take 0, 1, 3, 5, 8, 10 and 12 ml of the working standard sulphate solution in separate 25 ml of volumetric flasks and proceed as described above and run a blank simultaneously with the same reagents.

Observations and Calculation

The same as in the case of turbiditymetric method where given.

2.1.5.3 Ammonium Acetate-acetic Acid Extractable S

Principle

The method involves extracting the soil with ammonium acetate-acetic acid buffer solution. Acetate ions displace the sulphate ions from the adsorbed phase and from ammonium sulphate on combination. Thus the extract is composed of soluble sulphate plus a fraction of the adsorbed sulphate.

The aliquot of the extract is treated with barium chloride resulting in the precipitation of sulphate as barium sulphate. The extent of turbidity due to fine suspension of barium sulphate can be measured in the nephelometer (working on the principle of scattered light) using a standard curve obtained from barium sulphate, which causes opalescence of the solution. But within the range of low concentration, spectrophotometer or even colorimeter can work fairly satisfactorily. Measuring the degree of opalescence nephelometrically, the standard curve is drawn, which expresses the concentration as a function of the degree of opalescence of the solution analysed.

Apparatus

Photoelectric colorimeter or nephelometer,

Whatman No. 42 filter paper

Erlenmeyer flask; buchner funnel

Suction pump

Pipette

Volumetric flask.

Reagents

Extracting Solution: Dissolve 39 g of ammonium acetate in 1 litre of 0.25 N acetic acid (CH_3COOH).

Darco G-60 (activated charcoal): Wash the charcoal with extracting solution until it is free of sulphate.

Acid 'Seed' Solution: 6N HCl containg 20 ppm of S and K_2SO_4

Barium Chloride (BaCl_2. 2H_2O): 30-60 mesh crystals

Standard Sulphate Solution: Dissolve 0.5434 g of reagent grade K_2SO_4 in extracting solution in 1 litre volumetric flask and finally make up the volume to 1 litre with extracting solution. This solution contains 100 ppm S.

Procedure

☆ Shake 10 g of soil with 25 ml of the extracting solution in a 50 ml Erlenmeyer flask for 30 minutes.

☆ Add 0.25 g of Darco G-60 and resume the shaking for 3 minutes. Filter the soil suspension through Whatman No. 42 filter paper under suction.

☆ Pipette 10 ml of the filtrate into a 25 ml volumetric flask and add 1 ml of acid 'seed' solution, swirl the solution and add 0.5 g of $BaCl_2. 2H_2O$ crystals. Allow the mixture to stand for 1 minute and then swirl the solution in the flask frequently until the crystals are dissolved.

☆ Within the time span of 2 to 8 minutes.

☆ After the crystals have dissolved read the light transmittance or absorbance on a colorimeter or nephelometer at 420 nm using blue filter and find out the sulphate concentration from the standard curve.

☆ Simultaneously a blank must be run regardless of the purity of the chemicals and filter paper.

Observations and Calculation

☆ Weight of the soil taken = 10 gm

☆ Volume of the extract made = 25 ml

☆ Volume of the aliquot used = 10 ml

☆ Final volume made = 25 ml

☆ Transmittance (per cent T) as read for the colorimeter = T (say)

☆ ppm of S in the test solution as read from the standard curve = C (say)

☆ First dilution = 25/10 = 2.5 times

☆ Second dilution = 25/10 = 2.5 times

☆ Therefore, total dilution = 2.5 x 2.5 = 6.25 times

☆ Now, available S (ppm) in the soil = C x 6.25

☆ Hence, available S (kg/ha) in the soil = C x 6.25 x 2.24

2.1.5.4 Colorimetric Method for Determination of Available S Using Barium Chromate (Palaskar *et al.*, 1981)

Principle

The principle of the method is based on the extraction of S in 0.15 per cent calcium chloride in 1: 5 soil to solution ratio, using 10-20 g of soil with 30 minutes shaking (Williams and Steinbergs, 1959). Sulphur is estimated in a suitable aliquot, using barium chromate which reacts with sulphate in solution and precipitates the latter as $BaSO_4$, thereby releasing equivalent quantity of chromate ion, which is measured colorimerically.

The method can work well for a number of extractants other than those containing phosphate because when the reaction of the medium is changed from acidic to alkaline for removing unreacted barium chromate, barium phosphate also gets precipitated thus affecting the final colour intensity.

Apparatus

Photoelectric colorimeter

Whatman No. 42 filter paper

Conical flask

Volumetric flask

Pipette

Reagents

Barium Chromate Solution

Hydrochloric Acid (0.1 N)

Ammonium Hydroxide (5 N)

Standard Sulphate Solution: Prepare 100 ppm S AR grade from K_2SO_4 as described in the preceeding methods.

Procedure

☆ Weigh 10 g soil in a 100 ml conical flask and add 50 ml of 0.15 per cent $CaCl_2$. $2H_2O$ solution.

☆ Shake for 30 minutes and filter the contents through Whatman No. 42 filter paper.

☆ Pipette out 5 ml of barium chromate solution into a 100 ml volumetric flask, and add 1.2 ml of 5 N NH_4OH solution as to reduce the free aidity to about 0.05 N. No turbidity should appear.

☆ Transfer a measured quantity of the soil extract so that the sulphate sulphur content in the flask falls below 2000 μg.

☆ After shaking and standing for about half-an-hour, add 1 ml of NH_4OH to precipitated the unreacted chromate. Make up the volume to 100 ml with distilled water.

☆ Stopper flask and invert it 2-3 times. Filter the content through Whatman No. 42 filter paper, rejecting the first few ml of the filtrate.

☆ Measure the intensity of the yellow colour on a photoelectric colorimeter at 420 nm using blue filter

Note

It is necessary that the test solution should have a pH around 2 to 3 before barium chromate is added. In the case of alkaline extract, the pH has to be brought down with the required amount of 6N HCl. Since in the presence of phosphate an interference is likely and it can be precipitated with ferric salt ($FeCl_3$) and ammonia during the removal of precipitated barium sulphate and chromate. If excessive amount of phosphate is present in the test solution, turbidimetric estimation is recommended).

Preparation of Standard Curve

Take 0, 4, 8, 16 and 20 ml into 100 ml of 100 ppm SO_4^{2-} sulphur solution in separate volumetric flasks and read the colour intensity as per the procedure, mentioned above. Blank correction must be made.

Observations and Calculation

☆ Weight of the soil taken = 10 gm

☆ Volume of extractant added = 50 ml

☆ Volume of aliquot of the extract used = V ml

☆ Volume made = 100 ml

☆ Transmittance (per cent T) as read from the colorimeter = T (say)

☆ ppm of S from the standard curve against T = C (say)

☆ First dilution = 50/10 = 5 times

☆ Second dilution = 100/V times

☆ Total dilution = 5 x (100/V) = 500/V times

☆ Now, available S (ppm) in the soil = C x (500/V)

☆ Hence, available S (kg/ha) in the soil = C x (500/V) x 2.24

2.1.5.5 Determination of Exchangeable Calcium and Magnesium

Exchangeable calcium and magnesium can be determined in ammonium acetate extract by titration with ethyl diamine tatra acetic acid. Calcium, magnesium and a number of other ions forms stable complexes with ethyl diamine tatra acetic acid (EDTA) disodium salt. Cu, Zn, Fe, and Mn may interfere in the estimation of Ca and Mg, if present in appreciable amounts. Their interference is prevented by the use of 2 per cent NaCN solution or carbomate. A known volume of the solution is titrated with standard versenate (N/100) using erychrome black T chloride and ammonium purpurate at the end colour is changed from wine red to blue.

Apparatus

Vacuum pump

Hot plate

Beakers

Pipettes

Burette

Conical flask

Reagents

Buffer Solution: Dissolve 67.09 of ammonium chloride in 570 ml of conc. ammonium hydroxide and make the volume to 1 liter with distilled water.

Sodium Hydroxide (10 per cent): Dissolve 160 g of NaOH in 1 liter of water.

Standard Calcium Solution (0.01 N): Dissolve 0.50 g of pure $CaCO_3$ in 10 ml of approximately 3N (1+3) HCl and dilute to 1 liters with distilled water.

Eryochrome Black T (EBT): Dissolve 0.5 g of EBT and 4.5 g of hydroxyl amine hydrochloride in 100 ml of 95 per cent ethanol.

Ammonium Purpurate Indicator: Mix 0.50 g of ammonia purpurate with 100 g of powedered K_2SO_4.

Ethyl Diamine Tatra Acctic Acid (0.01 N): Dissolve 2.00 g of ethyl diamine tatra acctic acid and 0.05 g of magnesium chloride hexahydrate in water and dilute to a 1 liter with distilled water. Standardize the solution against standard calcium chloride solution

Procedure

Pretreatment of Soil Extracts

★ Ammonium acetate and dispersed organic matter when present in appreciable amount must be removed from soil extracts prior to titration with versenate.

★ Evaporation of an aliquot of the soil extract to dryness followed by treatment with aquaregia (3:1

conc. HCl + conc. HNO_3) and evaporation to dryness usually for the removal of ammonium acetate and organic matter. Very dark coloured soil may require additional treatment with aquaregia.

☆ Dissolve the residue in a quantity of water equal to the original volume of the aliquot taken for treatment.

Calcium

☆ Pipette 10 ml aliquot into a porcelain dish and dilute to 25 ml with distilled water

☆ Add 5 drops of NaOH (4 N) and 50 mg ammonium purpurate indicator.

☆ Titrate with EDTA. The colour change is from orange red to lavender or purple.

☆ Run a blank in the same fashion

Determination of Calcium + Magnesium

☆ Pipette 10 ml of aliquot into a conical flask and dilute to 25 ml with distilled water.

☆ Add 0.5 ml of buffer solution and 3-4 drops EBT indicator.

☆ Titrate with EDTA at the end point colour change from wine red to blue or green.

Calculation

Me/100 g soil of Ca + Mg.

$$= \frac{\text{ml of EDTA used x Normality of EDTA x volume of extract}}{\text{Wt. of soil}}$$

2.1.6 Determination of Exchangeable Sodium

Principle

Exchangeable sodium is determined by using ammonium acetate solution through flamephotometer. When a solution of a salt is sprayed into the frame it excites the atoms to higher energy level. When the electrons return back they emit radiations. The intensity of these emitted radiations is proportional to the concentration of the particular elements which is measured by flamephotometer.

Apparatus

Flame photometer

Volumetric flasks

Erlenmayer flasks

Pipettes

Reagents

Ammonium Acetate Solution (1 N): Dissolve 77.09 g of Ammonium acetate (NH_4OAC) in 1 litre of distilled water.

Standard Sodium Solution: Dissolve 5.845 g of NaCl in 1 N ammonium acetate solution and make the volume to 1 liter with distilled water. It will give 100 ppm Na.

Working Sodium Solution: Take 0, 1, 2,5, 5.0, 2.5 and 10 ml of 100 ppm Na solution in 100 ml volumetric flask and make the volume with 1 normal ammonium acetate solution to the mark . This will give 0, 1, 2, 7, 5 and 10 ppm sodium solution.

Procedure

☆ Take 5 g of soil sample in 150 ml erlenmayer flask and add 25 ml neutral normal ammonium acetate solution

☆ Shake on reciprocating shaker for 5 minutes and filter through Whatman No. 1

☆ Make the volume to 100 ml by ammonium acetate solution.

☆ Determine the Na with the help of flamephotometer using Na filter

Calculation

$$\text{Ammonium acetate extracted Na (ml/100 g)} = \frac{\text{Na conc. of extract in (me/1000 g)} \times 100}{\text{Weight of Soil}} \times \frac{\text{Volume of extract (ml)}}{1000}$$

2.1.7 Determination of Available Iron, Manganese, Copper, Zinc (DTPA extractable) by Atomic Absorption Spectrophotometer

The principle of atomic absorption and its application to the analysis of metals was described in 1955 by the Australian Physicist Alan Walsh. It includes all metals and semi-metals but excludes such elements as sulphur, halogens and other gases whose resonance wavelength lies in the atmospheric absorption region. Methods exist for indirect determination of these elements by combining them with a metal and then analysing for the metal. While the principle of atomic absorption spectrophotometer theoretically holds true for all elements the application of the method has been limited mainly to metals whole resonance lines lie in the ultraviolet and the visible regions of the spectra.

Principle

The method consists of shaking few grams of soil with a buffered solution, containing DTPA (diethylene triamine penta-acetic acid). This chemical acts as a mild chelating agent, which extracts the easily soluble zinc, iron, copper and manganese. The extracting solution is buffered at pH 7.3 with triethnolamine (TEA), and in addition, includes calcium chloride to prevent the dissolution of calcium carbonate. These conditions permit the right amount of zinc, iron, copper and manganese to be dissolved and $CaCl_2$ is to stabilize the pH of the extractant.

The dissolved elements are measured by the atomic absoption spectrophotometer, wherein the extracted sample is coverted first into an atomic vapor, usually by a flame, and irradiated by the metals being searched. The absorption of the light by the atomized samples is related to the concentration of the metal.

The narrow emission lines, which are to be absorbed by the sample, are, generally, provided by a hallow cathoge lamp, filled with neon or argon at a low pressure, which has a cathode lamp. Such lamp emits only the spectrum of desired element together with that of the filler gas. In the atomizer burner system, the solution to be analysed is drawn up by a capillary and coverted by means of a stream of compressed air into a fine spray, which, after condensation of larger droplets is mixed with acetylene, and burned in a long flame at a stainless steel burner. The light from the lamp, after traversing the flame, enters into monochromator, which has been set at a wavelength of the resonance line of the metal to be determined and then falls on the photomultiplier tube. The light is amplified by an A.C. amplifier whose output is read on read out device.

Apparatus

Atomic absorption spectrophotometer

pH meter

Horizontal shaker

Conical flasks

Volumetric flasks

Pipettes

Whatman No. 42 filter paper.

Reagents

Standard Solution for Zn, Cu, Fe and Mn: Prepare 100 ppm stock solution of each element as under:

Element	Atomic Weight	Salt	Mol. Weight	Weight of Salt g (for 100 ppm)
Zn	65.38	$ZnSO_4.7H_2O$	287.56	0.4398
Cu	63.54	$CuSO_4.5H_2O$	249.69	0.3929
Mn	54.94	$MnCl_2. 4H_2O$	197.91	0.3602
Fe	55.85	$FeSO_4.7H_2O$	278.02	0.4977

From the stock solution, prepare at least 10 working standards of each element with a range from 0 to 3.0 ppm for Zn and Cu, and from 0 to 20 ppm for Fe and Mn. Working standards should be prepared in the same matric as used for the extraction in soil samples. Measure the element in the extract by atomic absorption spectrophotometer first running a series of standards of the known element, then, analyzing the prepared samples.

DTPA Extracting Solution: To prepare 10 litre of this solution, dissolve 149.2 ml of TEA, 19.67 g of DTPA and 14.7 g of $CaCl_2. 2H_2O$ in approximately 200 ml of deionized water. Allow sufficient time for the DTPA to dissolve and dilute to 9 litre. Adjust the pH to 7.3 ± 0.05 with 1: 1 HCl while stirring and dilute to 10 litre.

The DTPA test is a nonequilibrium extraction. Therefore, factors, *viz.*, shaking time, shaking speed and the shape of the extraction vessel influence the quantity of metals extracted. These factors must be standardized, in each laboratory, or else the critical level, for each of the micronutrients, will be affected. Deviations from these standards will require recalibration of the test with plant growth response. If analysis is not completed within a day or two, refrigeration may be required to retard microbial growth.

Operating Parameters for Fe, Mn, Cu and Zn

According to the American Public health Association, (1985) of these parameters are listed below:

Parameters	Fe	Mn	Cu	Zn
Optimum conc. Range ($\mu g\ ml^{-1}$)	0.3-10	0.1-10	0.2-10	0.05-02
Sensitivity ($\mu g\ ml^{-1}$)	0.12	0.05	0.10	0.02
Detection limit ($\mu g\ ml^{-1}$)	0.02	0.01	0.01	0.005
Wavelength (mm)	248.3	279.5	324.7	213.9

Procedure

☆ Weigh 10 g of soil and add 20 ml of the extractant.

☆ Shake continuously for 2 hours preferably on a horizontal shaker and filter through Whatman No. 42 filter paper. If the filtrate is cloudy, refilter as necessary.

☆ Prepare at least 10 standards using DTPA as the matrix for each element with a range 0 to 3 ppm for Zn and Cu, and from 0 to 20 ppm for Fe and Mn.

☆ First take readings for the standards and then measure the element from the filtrate by atomic absorption spectrophotometer.

☆ A blank solution (without soil) should be run to correct for contamination.

Observations and Calculation

☆ Weight of soil = 10 gm

☆ Volume of DTPA extract made = 20 ml

☆ Reading on the galvanometer = T (say)

☆ Concentration (ppm) as read from the standard curve against T (or sample) = C (say)

☆ Concentration in the blank solution = Cb (say)

☆ Dilution factor = 20/10 = 2 times

☆ Now, available (ppm) heavy metal in the soil = (C-Cb) x 2

☆ Available (kg/ha)heavy metal in the soil = (C-Cb) x 2 x 2.24

2.1.8 Determination of Available Zinc

2.1.8.1 Ammonium-Acetate-Dithizone Extraction Method

Principle

Zn is extracted from soil by 1 N of ammonium acetate buffer in the presence of dithizone which forms complexes with zinc. These complexs are extracted by CCl_4 between pH. 8-10. Zn is determined colorimetrically.

Apparatus

Colorimeter

Centrifuge

Vertical shaker

Separatory funnel

Pipettes

Volumetric flasks

Reagents

Carbamate Solution: Dissolve 0.2 g in 100 ml of Zn free water.

Dithizone Solution in CCl_4 (0.1 per cent): Dissolve 0.2 g of carbon tetra chloride in 1 litre of distilled water. Shake frequently in a 4-1 separatory funnel for 15 minutes. Add 2 ml of Zn free 0.02 N ammonium hydroxide and shake to transfer dithizone to aqueous phase. Discard the carbon tetrachloride (light green colour) and since the aqueous wash with several 100 ml portions of CCl_4. Add 500 ml CCl_4 and 50 ml of Zn free HCl. Shake to transfer dithizone to CCl_4. Dilute CCl_4 dithizone wash to 2 litres keep in a glass stored pyrex bottle stored in a refrigerator.

Ammonium Acetate Buffer: Dissolved 77.0 g of ammonium acetate in adjust pH to 7.0 with Zn free NH_4OH pored in a separatory funnel with dithizone and CCl_4, discarding the organic phase until it no longer changes colour. Remaining dithizone dissolved in aqueous or phase by repeated extractions with CCl_4.

Ammonium Citrate Buffer (0.4 M): Add 90 g of dibasic ammonium citrate buffer to distilled water and make the volume to 1 litre with distilled water. Add conc. NH_4OH until the pH reaches to 8.5. Purify with 100 ml of dithizone and 100 ml of CCl_4.

Hydrochloric Acid (1 N): Distilled 6 N of HCl to IN HCl with Zn free water.

Ammonium Hydroxide (IN NH$_4$OH): Distill concentrate NH$_4$OH into Zn free water in Pyrex container immersed in an ice bath. Zn free NH$_4$OH may also be prepared by collecting anhydrous NH$_3$ in a Pyrex container of Zn free water.

Zn Free Water: Red still water in Pyrex glass or pass distilled water through an ion exchange column.

Standard Zn Solution (100 ppm): Dissolve 0.1 g of pure Zn in 50 ml of 0.02 N H$_2$SO$_4$. Dilute to 1 litre with distilled water.

Procedure

☆ Pipette 25 ml of ammonium acetate and 25 ml of dithizone CC1$_4$ solution into a 125 ml separatory funnel.

☆ Add 2.5 g of soil and shake on a vertical shaker for 1 hour

☆ Drain the soil CCl$_4$ suspension into a 40 ml centrifuge tube. Contribute to develop a continuous CC1$_4$ phase.

☆ Take 10 ml aliquot and soil layers and the tip must be kept away from the walls of the tube where soil and water may be contacted and rinse the adhering soil and water from the tip, using a wash bottle and adjust to the mark and rinse again.

☆ Add the aliquot to the clean separatory funnel from which it originally came and which was cleaned by rinsing with water while samples were being centrifuged. Add 50 ml of 0.02 N HCl and shake for 3 minutes.

☆ Discard the CCl_4 phase.

☆ Rinse the aqueous phase twice with CCl_4, shaking by hand.

☆ Add 1 drop of phenolphthalein indicator 5 ml of ammonium citrate buffer, 1.1 ml of 1 N NH_4OH, 5 ml of carbomate and 10 ml of dithizone reagent.

☆ Shake and transfer the organic phase into another separatory funnel.

☆ Add 25 ml of conc. 1 N NH_4OH and extract again for 3 minutes. Take 5 ml aliquot and dilute with CCl_4 to 25 ml.

☆ Measure the intensity by colorimeter at 535 nm using green filter

Observations and Calculation

☆ Weight of soil taken = 2.5 g

☆ Volume of dithizone extract prepared = 10 ml

☆ Volume of dithizone extract taken for colour development = 5 ml

☆ Volume made upto with CCl_4 = 25 ml

☆ Transmittance (per cent) as read from the colorimeter = T (say)

☆ Conc. Of Zn (ppm) from the standard curve against T = C (say)

☆ First dilution: 10/2.5 = 4 times

☆ Second diltion: 25/2 = 5 times

☆ Total dilution = 4 x 5 = 20 times

☆ Available Zn (ppm) in the soil = C x 20

☆ Thus, available Zn (kg/ha) in the soil = C x 20 x 2.24

2.1.9 Determination of Available Manganese

Principle

The soil is extracted with neutral normal ammonium acetate solution and the extracted Mn is determined colorimetrically based on the oxidation of Mn^{2+} to MnO_4^- in strong acid conditions.

Apparatus

Colorimeter

Mechanical shaker

Vacuum pump

Conical flask

Volumetric flask

Reagents

Ammonium Dihydrogen Phosphate: Dissolve 115 g of $NH_4H_2PO_4$ in water and dilute to 1 litre.

Silver Nitrate: Dissolve 2 g $AgNO_3$ in 100 ml of water and keep in a dark bottle.

Potassium Metaperiodate

Standard Manganese Solution (100 ppm): Dissolve 0.3076 g pure $MnSO_4 \bullet 7H_2O$ in 1 liter of distilled water

Hydrochloric Acid (HCl)

Nitric Acid (HNO_3)

Perchloric Acid ($HClO_4$)

Phosphoric Acid (H_3PO_4, 60 per cent)

Potassium Peroxidate (KIO_4, 85 per cent)

Procedure

☆ Weigh 10 g of soil in 500 ml flask, add 250 ml of

Ammonium dihydrogen solution shake for 30 minutes and filter.

☆ Take 25 ml aliquot in 100 ml volumetric flask, To remove chlorides add 5 ml conc. H_2SO_4 and keep on a hot plate.

☆ Add 5 ml of each of nitric and phosphoric acids and dilute to approximately 80 ml.

☆ Add 1 ml of $AgNO_3$ solution and 0.3 g of potassium peroxidate powder. Heat to boiling and keep for 10 minutes

☆ Cool, dilute to 100 ml with distilled water.

☆ Measure the intensity of colour at 525 nm by spectrophotometer.

Observations and Calculation

☆ Weight of soil = 10 g

☆ Volume of extract = 250 ml

☆ Volume of aliquot taken = V_1 ml

☆ Volume made upto while developing colour = V_4 ml

☆ Transmittance (per cent) as read from the colorimeter = T

☆ Concentration (ppm) of Mn from the standard curve = C

☆ Available Mn (ppm) = $C \times V_4 \times 250/V_1 \times 1/10$

$$= C \times V_4 \times 10/V_1 = A \text{ (Say)}$$

☆ Available Mn (kg/ha) = A x 2.24

2.1.10 Determination of Available Copper

The copper content of soils ranges from 10-80 ppm

depending upon the nature of parent material. The solumetry of copper is highly pH dependent and at low ambient level its diffusion to plant roots may be the limiting factor in alkalie and calcarious soils.

Acid Ammonium Acetate Extraction Method

Principle

The soil is treated with ammonium acetate buffer followed by acid treatment, to oxidize organic matter. A known volume of aliquot is then treated with hydroxyl amine hydrocholoride for ensuring complete reduction of copper. The cuprous copper is then estimated colorimetrically.

Apparatus

Colorimeter

Sepratory funnel

Volumetric flask

Pippets

Reagents

Nitric Acid conc. (HNO$_3$)

Perchloric Acid (60 per cent)

Hydroxylamine Hydrochloride Solution

2,2,-biquinoline Solution: Dissolve 0.2 g of 2,2-biquinoline in 800 ml of iso-amyl alcohol and dilute to 1 litre with iso-amyl alcohol, and store it in an amber colour bottle.

Ammonium Acetate Buffer: Add 1,270 ml of conc. NH$_4$OH and 805 ml of CH$_3$COOH (glacial) to 18 liters of redistilled water. Adjust the pH to 4.8 with NH$_4$OH or CH$_3$COOH.

Standard Copper Solution: Dissolve 0.3928 g $CuSO_4.5H_2O$ in distilled water containing 2 ml of conc. H_2SO_4, and dilute to 500 ml with distilled water. The solution contains 200 ppm of Cu. Dilute 50 ml (200 ppm) to 1000 ml with redistilled water to obtain 10 ppm Cu solution.

Procedure

☆ Weigh 50 g of air-dry soil in a stoppered bottle

☆ Add 100 ml of buffered NHOAc, and shake for 1 hour

☆ Filter the suspension through Whatman No. 42 filter paper.

☆ Pipette 50 ml of the filtrate into a beaker, and evaporate the solution to dryness on a hot plate.

☆ Add 5 ml of conc. HNO_3 and 2 ml of $HClO_4$, and evaporate the solution slowly to dryness.

☆ Dissolve the residue in 50 ml of redistilled water

☆ Pipette 10 ml of the sample solution into a separating funnel, and mix thoroughly with 25 ml of NH_4OAC buffer and 0.5 g of hydroxylamine hydrochloride.

☆ Allow the phases to separate and discard the aqueous phase.

☆ Transfer the iso-amyl alcohol to a centrifuge tube, and centrifuge at 1,200 rpm for 5 minutes.

☆ Prepare standards containing 0, 1, 2, 3,4, 5 µg of Cu in the manner described for the sample aliquots.

☆ Place the tubes in the spectronic 20, and measure percentage transmittance at 540 mµ light maximum. Construct a calibration curve for cuprous complex. From this curve and the sample dilution, calculate the available Cu content of the soil after subtracting Cu present in the reagents.

Observations and Calculation

☆ Weight of soil = 50 gm

☆ Volume of extract = 100 ml

☆ Volume of aliquot taken = 50 ml

☆ volume made up following heat and acid treatments = 50 ml

☆ Volume for sample solution = 10 ml

☆ Volume made = 50 ml

☆ Transmittance (per cent) value = T (say)

☆ Concentration (ppm) from the curve against T = C (say)

Available Cu (ppm) = C x 50 x 50/10 x 100/50 x 1/50

= C x 10

Available Cu (kg/ha) = C x 10 x 2.24

The copper may be determined at 520 nm by spectrophotometer

2.1.11 Determination of Available Iron

2.1.11.1 Colorimetric Extraction Method

Principle

Soil is extracted with 1 N-NH_4OAc (pH 4.8). Iron is first reduced with hydroxypamine hydrochloride and then allowed to react with orthophenanthroline. A stable red colour complex is formed by iron and orthophenoanthroline due to formation of tri (1,10) Phenoanthroline ferrous ion ($Fe[C_{12}H10N_2]_3^{+2}$. NH_4OAc extract of most soils are slightly coloured because of organic matter, a correction is made for this colour with each soil. This is done by setting the photometer to indicate 100 per cent light transmission using a sample of extract to which all reagent except orthophenthroline have been added.

Apparatus

Photoelectric colorimeter

pH meter

Vacuum pump

Water bath

Pipette

Beakers

Reagents

Ammonium Acetate: (1 N, NH$_4$OAc): Take 102 ml of glacial acetic acid and 70 ml of conc. NH$_4$OH to 750 ml of water. Adjust the pH to 4.8 by adding CH$_3$COOH or NH$_4$OH and dilute the solution to one litre.

Hydroxylamine Hydrochloride (10 per cent): Add 90 ml of distilled water to 10 g of hydroxylamine hydrochloride.

Orthophenanthroline Reagent (0.4 per cent): Dissolve 0.3 g of orthophenan-throline monohydrate in water by heating the mixture to 80°C . Cool the solution and add water to make a volume of 100 ml.

Standard Iron Solution (100 ppm): Dissolve 0.7022 g ferrous ammonium sulphate in distilled water and add 50 ml of 0.6 N HCl and make the volume to 1 liter with distilled water.

Working Iron Standard: Take 0, 0,1, 0.2, 0.4, 0.6, 0.8, 1.0, 1.6 and 2 ml 100 ppm Fe solution. Add 2 ml each hydroxylamine hydrochloride and orthophenanthroline read and the colour intensity at 510 nm by colorimeter.

Procedure

☆ Place 2.5 g of soil in a flask and add 50 ml of NH_4OAc reagent. Stopper the flask and shake it for 30 minutes on a mechanical shaker.

☆ Filter or centrifuge the suspension, and pipette a 10 ml partion of the extract into each of two photometer tubes.

☆ To the first tube, add 2 ml of 10 per cent hydroxylamine hydrochloride reagent. Mix the solution and add 2 ml of orthophenthroline reagent.

☆ To the second, add 2 ml of 10 per cent hydroxylamine hydrochloride and 2 ml of water.

☆ Read the colour intensity at 510 nm by spectrophotometer

Observations and Calculation

☆ Weight of soil = 25 g

☆ Volume of extract = 100 ml

☆ Volume of aliquot taken = V_1 ml

☆ Final volume made = V_2 ml

☆ Per cent transmittance = T

☆ Available Fe (ppm) from standard curve = C

☆ First dilution = 100/25 = 4

☆ Second dilution = V_2/V_1

☆ Total dilution = $V_2/V_1 \times 4$

☆ Available Fe (ppm) = $C \times V_2/V_1 \times 4$

☆ Available Fe (kg/ha) = $C \times V_2/V_1 \times 4 \times 2.24$

2.1.12 Determination of Available Molybdenum

The extractants used for estimation of available molybdenum in soil are acid ammonium oxalate, hot water and one molar neutral ammonium acetate solution. The acid ammonium oxalate extraction method is considered to be the best for available molybdenum.

Principle

Ammonium oxalate produces oxalate ions which may replaced the molybdnum on the soil colloids and clay particles. The replaced molybdnum is determined by the colorimetric method.

Reagent

Acid Ammonium Oxalate Solution (0.2 M): Dissolve 24.9 g of ammonium oxalate and 12.6 g of oxalic acid in distilled water. Dilute to approximately 900 ml with distilled water adjust the pH to 3.3 and finally make up the volume to 1 litre with distilled water. .

Hydrochloric Acid Feric Chloride Buffer Solution: Add 0.5 g of Fe Cl_3 $6H_2O$ to 560 ml of conc. HCl and dilute to 1 litre.

Sodium Thiocyanate: Dissolve 30 g sodium Thiocyanate in water and dilute to 100 ml with distilled water.

Stannous Chloride: Dissolve 40 g of $SnCl_2$ in 20 ml of 6 N HCl. Add water to dissolve and dilute to 100 ml with distilled water.

Standard Molybdnum Solution: Dissolve 75 g of MoO_3 in a slight excess of NaOH and dilute to 300 ml with distilled water. Make slightly acidic with HCl and dilute to 500 ml dilute 20 ml of this solution to 1 litre to get 2 ppm Mo solution

Apparatus

    Spectrophotometer

    Muffle furnace

    Mechanical shaker

    Water bath

    Balance

    Separatory funnel.

    Conical flask

    Erlenmeyer flasks

    Beakers

Procedure

☆ Weigh 25 g soil into a 500 ml Erlenmeyer flask and add 250 ml of the acid ammonium oxalate solution.

☆ Shake overnight; filter and evaporate 200 ml of the filtrate to dryness in 100 ml pyrex beaker.

☆ Heat in a muffle furnace at 450°C for 4 hrs to destroy organic matter. Take the residue in 10 ml of distilled water, add 10 ml of conc. HCl and filter.

☆ Wash with distilled water to make the volume less than 30 ml.

☆ Transfer to a separatory funnel. Add 10 ml of the hydrochloric acid-ferric chloride reagent and make the volume to 45 ml with distilled water.

☆ Add 3 ml of isoamyl alcohol-carbon tetrachloride extractant. Shake for 2 minutes.

☆ Allow 10 minutes to separate the organic phase and discard the extractant.

☆ Add 1.0 ml of the sodium thiocynate solution and mix.

☆ Add 1.0 ml of the $SnCl_2$ reagent and mix again. Add 1 ml of the extractant.

☆ Shake for 2 minutes, releasing pressure as necessary.

☆ Wash the tip of the funnel with distilled water and dry the tip with vacuum through tipped tube. Open the stopcock of the inverted funnel and dry the stop cock base with vacuum.

☆ Shake the funnel quickly and allow the organic phase to separate.

☆ Discard 1 or 2 drops of the extractant and transfer the remainder to the absorption cell. Measure the colour intensity at 470 nm using water as a reference.

Observations and Calculation

☆ Weight of soil = 25 g

☆ Volume of the ammonium oxalate extract made upto = 250 ml

☆ Volume of extract = V ml

☆ Volume of aliquot made upto following various treatments = V_1 ml

☆ Transmittance (per cent T) as read from the colorimeter = T

☆ Concentration (ppm) of molybdenum = C

☆ First dilution = (250/25)

☆ Second dilution = (V_1/V)

☆ Total solution = $10 \times (V_1/V)$

☆ Now, accessible Mo content of soil (ppm) = C x 10 x (V_1/V)

☆ Hence, accessible Mo content of soil (kg/ha) = C x 10 x (V_1/V) x 2.24

2.1.13 Determination of Available Boron

Water soluble boron is the available form of boron. It is extracted from the soil by water suspension. In the extract boron can be analysed by colorimetric such as Curcumin, Azomethine–H and most recently by inductively coupled plasma atomic emission spectrophotometry.

2.1.13.1 Curcumin Method

Principle

The curcumin reagent 1-7 bis (4-hydroxy-3 methoxyphenyl)-1, 6-heptadiene-3, 5-dione, is evaporated to dryness in an acid solution containing boric acid, a red colour is developed and is intensified by oxalic acid. The product is called rosecyanine and is soluble in ethyl alcohol. The intensity of the colour is proportional to the boron content.

Apparatus

    Boron free glassware

    Conical flask

    Volumetric flask

    Pipette

    Water bath

Reagents

    Ethyl Alcohol (95 per cent)

Curcumin Oxalic Acid: Dissolve 0.04 g of finely ground curcumin and 5 g of oxalic acid in 100 ml of 95 per cent ethyl alcohol.

Calcium Hydroxide Suspension: Dissolve 0.4 g of Ca $(OH)_2$ in 100 ml of distilled water.

Hydrochloric Acid (2.0 N)

Phenolphthalein Indicator: Dissolve 0.05 g of phenolphthalein in 50 ml of 95 per cent ethyl alcohol and 50 ml of water.

Standard Boron Solution (100 ppm): Dissolve 0.05 g of boric acid in distilled water and dilute to 1 liter.

Working Boron Solution: Dilute 10 ml of stock solution to 1 liter giving 100 ppm boron (B) concentration.

Barium Chloride Solution (10 per cent).

Preparation of a Calibration Curve: Take 0, 1, 2, 3, 4 and 5 ml of standard boron (B) solution into separate evaporating dishes

Add 5 ml of Ca $(OH)_2$ suspension and evaporate the contents to dryness. Allow the dish to cool, and then add one drop of phenolphthalein indicator.

Add 2.5 N HCl drop by drop until the colour disappears

Add 0.5 ml of the acid and add 4 ml of curcumim-oxalic acid solution and rotate the dish.

Read the colour intensity at 540 nm using red filter by colorimeter

Procedure

☆ Weigh 20 g soil in 100 ml volumetric flask gently reflex with and add 40 ml of distilled water and 0.5 ml of 10 per cent $BaCl_2$ solution.

☆ Cool decent the suspension heat and centrifuge for 5 minutes at 1500 to 2000 rpm.

☆ Add 4 ml of the curcumin-oxalic acid solution and rotate.

☆ Place the dish on a water bath at $55 \pm 3^{\circ}C$. If a controlled heater bath is not available, regulate the temperature by a gas burner and allow for 15 minutes after the liquid has evaporated.

☆ After that decant the solution into a 25 ml volumetric flask, and add alcohol to make the volume to 25 ml.

☆ Measure the absorbance of the solution at 540 nm.

☆ Calculate the concentration of boron in the test solution from the standard curve.

Observations and Calculation

☆ Weight of soil = 20 g

☆ Volume of extract made upto = V ml

☆ Volume of aliquot = V_1 ml

☆ Final volume made = 25 ml

☆ Transmittance (per cent T) as read from the colorimeter = T

☆ Concentration of boron (ppm) = C

☆ First dilution = $(V/20)$

☆ Second dilution = $(25/V_1)$

☆ Total dilution = $V/20 \times 25/V_1$

☆ Available B in soil (ppm) = $C \times V/20 \times 25/V_1$

☆ Available B (kg/ha) = $C \times V/20 \times 25/V_1 \times 2.24$

Chapter 3

Testing for Edaphic Chemical Properties

3.1 Soil Texture

Soil texture describes the relative proportion by mass of mineral particles of soil called soil separates: Sand (2.0-0.5), silt (0.05-0.002 mm) and clay (less than 0.002 mm) as per USDA Textural Classification.

3.1.1 Determination of Texture

Soil texture is generally determined in the laboratory through mechanical analysis following either the international pipette or using Bouyoucos hydrometer method.

3.1.1.1 Hydrometer Method

Principle

The estimation of texture by mechanical analysis is based on two processes: dispersion and sedimentation which separate the soil into the three classes. Dispersion refers to the separation of individual soil particles and is achieved by chemical and mechanical means. Mechanical stirring disperses larger aggregates while chemicals dispersing (through sodium-hexa-metaphosphate) is required for small aggregated clay groups. The mixture of dispersed soil particles in water is called a soil suspension. Sedimentation refers to the setting rates of the dispersed particles in water which is a function of particle size and governed by Stoke's law which is given below:

$$V = \frac{2\ gr^2}{9\ n}\ (P_s - P_w)$$

in which,

V: Setting velocity (cm S^{-1})

r: Effective radius of the particle (cm)

g: Acceleration due to the gravity (cm S^{-1})

P_s: Density of the particle (g cm^{-3})

P_w: Density of water (g cm^{-3})

n: Kinetic viscosity of water ($cm^2\ S^{-1}$)

The time of sampling (t) is dictated by the sampling depth (or the point where the density is to be measured) and the equivalent diameter (d) of the particles by the following formula:

$$T = \frac{H}{V} = \frac{18\ n\ h}{(P_s - P_w)\ d^2}$$

For a fixed 'h' all the terms on the right hand side in the above equation are constant except 'd'.

A hydrometer is used to determine the density of suspension at any given time. Bouyoucos determined that after 40 seconds the hydrometer reading is not affected by size of the particles. It implies that whatever remains in the suspension is composed of silt and clay particles. After two hours, only the particles less than 0.002 *i.e.*, clay particles are left in the soil suspension.

Apparatus

Balance

Oven

Measuring cylinder

Stirring rod

Electrical stirrer with baffle cup

Thermometer

Hydrometer

Reagent

Sodium-hexa-metaphosphate (5 per cent) solution.

Procedure

☆ Weigh 50 g oven dried soil sample into a stirring cup. Fill half of the cup with distilled water and add 100 ml of 5 per cent sodium-hexa-metaphosphate solution.

☆ Place the cup on a stirrer and stir for 10 minutes

☆ Transfer the suspension into a setting cylinder and make the volume of suspension to the 1 litre with distilled water.

☆ The setting cylinder is closed by a tightly fitting rubber stopper and shaken vigorously back and forth and then placed on the table and the time recorded. Alternatively the solution can be stirred with a stirrer.

☆ Immediately the hydrometer is lowered gently inside the cylinder and the first reading is taken exactly 40 seconds after the shaking was stopped.

☆ The temperature of the suspension is also recorded. For each degree above 68°F add 0.2 to the hydrometer reading. For each degree below 68°F subtract 0.2 from the hydrometer reading. This is the corrected hydrometer reading. The correction factor is 0.36 for each degree above/below 20°C.

☆ The suspension is shaken vigorously and again placed on a table where it will not be disturbed. The hydrometer reading is now taken after two hours. Temperature correction is also done as described earlier.

☆ The sand, silt and clay composition is calculated and the texture determined using the USDA textural triangle (Figure 3.1).

A preliminary idea of soil texture in the field also can also be obtained by the Feel Method (Table 3.1).

The information generated from the mechanical analysis can also be used to calculate Dispersion and Erosion Ratio. These two properties are given by:

$$\text{Dispersion ratio} = \frac{\text{Amount of silt + clay obtained after water dispersion of the soil}}{\text{Total amount of silt + clay obtained after complete dispersion of same soil}}$$

Table 3.1: Determination of Soil Texture (Feel Method)

Textural Class	Feel of Fingers	Ball Formation	Stickiness	Ribbon Formation
Sand	Very gritty	Does not form ball	Does not form ball	No
Loamy sand	Very gritty	Forms very easily with broken ball.	Very little, stains fingers slightly.	No
Sandy Loam	Moderatlely gritty	Forms fairly firm ball but is easily broken.	Definitly stains fingers	No
Loam	Neither very gritty nor smooth.	Forms firm ball	–do–	No
Silt loam	Smooth on slick buttery feel.	–do–	–do–	Slight tendency to ribbon with blacky surface.
Clay loam	Slightly gritty feel	Moderately hard ball when dry.	–do–	Ribbons out on squeezing but ribbon breaks easily
Silty clay loam	Very smooth	–do–	–do–	Shows some flasking on ribbon surface similar to silt loam.
Clay	Very smooth	Forms hard ball, which when dry can not be crushed by fingers.	–do–	Squeezes out at right moisture into long (1"-3") ribbons

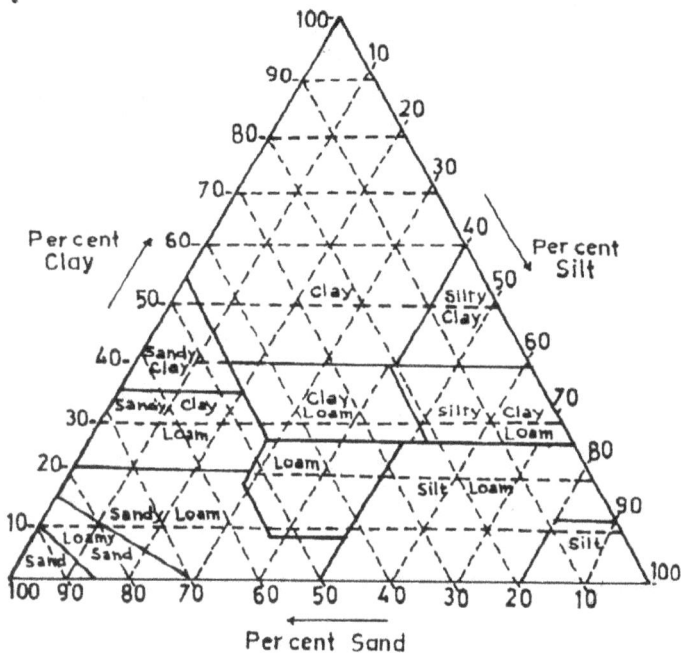

**Figure 3.1: Textural Chart Identifying
Soil Types by Sand, Silt and Clay Contents**

$$\text{Erosion ratio} = \frac{\text{Dispersion ratio}}{\text{[Total amount of Silt + Clay in soil]}/\text{[Moisture equivalent ratio (FC value)]}}$$

Formula for conversion of °F to °C   °C = {5 (°F–32)}/9

(since hydrometers are calibrated in °F)

3.1.1.2 International Pipette Method

Principle

The principle of particle–size analysis of soil requires complete dispersion of the soil mass initially. This is achieved by removing, from the soil, the cementing agents and the flocculating agents, in succession.

Removal of cementing agents is done by pretreating a given mass of dry soil with hydrogen peroxide to oxidise organic matter, followed by treatment with HCl to remove carbonates. Soil particles are then dispersed in a volume of standard NaOH and the particles are separated in the suspension by shearing action or turbulent mixing, using an electrical stirrer. The coarse sand fraction is then quantitatively removed from the dispersed soil by using a 0.02 is sieve.

The sedimentation procedure, in the method, involves the determination of the concentration of soil particles, by homogenising a one litre soil suspension, using the technique of pipette sampling at controlled depth of time. The concentration, at depth h and time t will be the concentration of particles of diffeent sizes, having velocity < h/t. Silt and clay are determined, in the suspension, based on their settling times. Following the removal of silt and clay the fine sand fraction is quantitatively transferred and its content determined. Various fractions are dried at 105°C and all results should be expressed as percentages of oven-dried weight.

Apparatus

> 500 ml beaker
>
> Hot water bath
>
> Rubber tipped glass rod
>
> Bottle
>
> Filter paper (Whatman No. 50)
>
> Electric stirrer

Rubber pestic

Sieve (70-mesh)

Measuring cylinder (1 liter cap)

Glass funnel

Thermometer

Pipette

10 m cap (preferably Robinson's pipette)

An iron rod attached to a circlar metallic disc or 5-6 cm in diameter, and about 0.15 cm in thickness, *i.e.*, a plunger,

Porcelain dishes or silica cruciblies

Balance

Oven

Reagents

Hydrogen Peroxide (6 per cent)

Hycrochloric acid (2 N)

Silver nitrate (N/10)

Sodium hydroxide (N/10) or NaHPO$_4$ (1N)

Phenolphthalein indicator

Procedure

(a) Treatment with Hydrogen Peroxide

☆ Take 20 g of air dry soil in a 500 ml beaker

☆ Add 50-60 ml of 6 per cent H$_2$O$_2$ and swirl well.

☆ Place the beaker over a hot water bath

☆ Continue digestion, till the reaction completely subsides. In case of frothing persists, add a second lot of 30-40 ml of H_2O_2 and see for effervescence, if any. Continue the step till the whole of the organic matter is oxidised.

☆ Cool and rinse its sides with a rubber pestle and with a jet of distilled water

(b) Treatment with Acid and Filtration

☆ Add 25 ml of 2 N HCl and shake to destroy $CaCO_3$ if the soil contains more than 2 per cent $CaCO_3$, more HCl should be added at the rate of approximately 250 ml for each per cent of $CaCO_3$.

☆ Dilute to 250 ml and thoroughly rub the soil with a rubber pestle. Allow the contents to react for about an hour, with intermittent shaking.

☆ Filter the contents through a buchner funnel under vacuum, using a Whatman No. 50 filter paper.

☆ Wash the soil with distilled water till the filtrate runs free of chloride (this can be tested with $AgNO_3$ solution).

(c) Dispersion and Separation of Coarse Sand

☆ Transfer the soil sample to a 500 ml beaker with a jet of distilled water, and make the volume to about 400-500 ml with distilled water.

☆ Add a few drops of phenolphthalein indicator.

☆ Add NaOH till the whole suspension shows a pink colour, indicating its alkaline reaction.

☆ Stir the content with an electric stirrer for 10 minutes

☆ Transfer the content to a 70 mesh (0.2 mm) sieve,

☆ Pour the suspension into the sieve, and with a jet of distilled water wash as much material as possible, through the sieve, until no more clay and silt remain on the sieve and the cylinder is about one half-full.

☆ Transfer the coarser particles, retained on the sieve, to a weighed dish to a constant weight. Record the weight.

(d) Separation of Silt and Clay

☆ Make up the volume of the suspension to one litre by adding more distilled water.

☆ Next day place the cylinder in a temperature controlled chamber to ensure minimum variation of temperature between the two samplings. Alternatively, measure the temperature of the suspension, find out the requisite time for sedimentation, corresponding to this temperature, for silt and clay determination.

☆ A little before the time of sampling stir the cotton with plunger by moving it up and down, about 20-25 times in one minutes, so as to ensure stability of the suspension.

☆ Remove the plunger gently and note down the time at which the plunger is taken out. This is the time of commencement of sedimentation.

☆ Insert the sampling pipette gently into the suspension and dip it to 10 cm depth from the surface of the suspension about 20 second before the time is up, and withdraw 10 ml suspension.

☆ Deliver the content of the pipette into a weighed dish and dry the sample at 105°C to a constant weight.

(e) Separation of Clay

☆ Stir the content of the cylinder again for one minute as before.

☆ Sample in the same way at a dept of 10 cm immediately after the expiry of the requisite time for clay.

☆ Transfer the suspension to a weighted dish, dry it at 105°C and record the weight.

(f) Separation of Fine Sand

☆ Decant the bulk of the suspension and transfer the sediment, containing the fine sand, to a 500 ml beaker with a jet of distilled water.

☆ Pour away the turbid suspension, each time the sediment is washed with distilled water.

☆ Continue the last step until the liquid no longer stands turbid.

☆ Remove quantitatively with a jet of distilled water, the particles, left in the beaker to a weighed dish and record its weight, after overdrying at 105°C to a constant value.

☆ Compare the fine sand, thus, obtained with that obtained by the difference method.

☆ Take a given amount of the same soil, and record its oven-dry weight for determination of water content of soil.

Observations and Calculation

☆ Weight of the soil taken = 20 g

☆ Water content of the soil or moisture per cent thus, oven-dry weight of the soil = 100 x 20 /100 + b = W gm (say)

☆ Weight of dish = W_1 gm

☆ Weight of dish + coarse sand = W_2 gm.

Therefore weight of coarse sand = W_2-W_1

Percentage of coarse sand pcs = $\dfrac{(W_2-W_1 \times 100)}{W}$

☆ Temperature of the suspension = T°C

☆ Time at which the plunger it taken out the cylinder = t_0

☆ Time at which the sample for silt + clay is to be collected = $t_0 + t_{sic}$ (hours, minutes, seconds)

☆ Time at which the sample for clay is to be collected = $t_0 + t_c$ (hours, minutes, seconds)

☆ Weight of dish = W_3 gm

☆ Volume of suspension taken for analysis = 10 cm

☆ Weight of dish and silt + clay = W_4 gm

☆ Weight of silt + clay = (W_4–W_3) gm thus

% silt + clay, Psic = $\dfrac{W_4 - W_3}{W} \times \dfrac{1000}{10} \times 100$

At $t_0 + t_c$

☆ Weight of dish = W_5 gm

☆ Volume of suspension taken for analysis = 10 ml

☆ Weight of dish + clay = W_6 gm

☆ Weight of clay alone = (W_6-W_5)

Therefore, percentage of clay, Pc

$$\frac{(W_6-W_5 \times 1000 \times 100)}{W \times 10}$$

Therefore, percentage silt Psi = Psic-Pc

☆ Weight of dish = W_7 g

☆ Weight of dish + fine sand = W_8 g

Thus, per cent of fine sand P_{fs}

1. Experimental = $\dfrac{W_8-W_7}{W} \times 100$

2. By subtraction = 100–(Pcs + Psic)

3.2 Determination of Bulk Density of Soil

Principle

Bulk density of soil (Pb) is the dry mass per unit volume of a given soil in its natural undisturbed condition and is given by:

$$Pb = \frac{MS}{Vb}$$

where,

MS is the mass of dry solid (g).

Vb is the volume of soil (cm^3).

Pb is bulk density of soil $(g\ cm^{-3})$.

It is required to convert soil water content determined on weight basis to volume basis which have more practical utility.

Since the bulk volume includes the actual volume occupied by the soil solids and pore volume, bulk density will, therefore, vary not only with the actual density of the solids but also more importantly with the packing of the soil practices. Hence, a given soil can have widely varying bulk density depending upon the state of compaction and aggregation. Tillage operations that loosen soils temporarily lower bulk density, while compaction raises bulk density. Natural soil forming process that increase aggregation reduce bulk density. Processes or operations like heavy machinery that destroy aggregation increase bulk density.

The bulk density of fine textured mineral soils ranges from 1.0 to 1.3 g cm^{-3} and that of sandy soils from 1.4 to 1.7 g cm^{-3}. Any of the following three methods can be employed to determine the bulk density.

3.2.1 Core Sampler Technique

Bulk density is generally obtained by removing a block of soil from the site under study allowing no compaction or crumbling. This can be accomplished by using any method that forces a straight-sided container of known volume into the soil without altering the area to be sampled. The container is then dug out, excess soil is trimmed away, dried and weighed and Pb calculated. This involves inserting a core of known dimensions into the soil and extracting a sample of known volume. In the absence of a core sampler a rigid shallow cylindrical tin (metallic) can also be used under field conditions.

3.2.2 Sand Pouring Technique

In this approach the soil sample is first extracted from a small pit dug at the site and then the volume of resulting cavity (space) is determined by pouring sand into it from a measuring cylinder and noting the volume of sand used. The oven dried mass of the soil sample extracted is determined gravimetrically.

3.2.3 Paraffin Cold Technique

A soil clod or aggregate is collected, coated with an impermeable material such as paraffin wax and its volume determined by fluid displacement. This is accomplished by dipping the clod tied with a string into a container of melted wax. Subsequently the oven dry mass of the clod is determined gravimetrically.

Apparatus

Balance

Oven

Core sampler assembly with cores

Spatula

Procedure

☆ The core sampler is assembled with the core placed inside after a thorough cleaning.

☆ For surface samples, prepare the soil surface by removing vegetation and loose soil and smoothen the soil surface. For samples from lower depth excavate the soil to the appropriate depth. (Layerwise bulk density samples are best taken when the soil profile pit studies are being carried out).

☆ Drive the core sampler into the soil with the hammer.

☆ Carefully dig out/remove the core sampler and dismantle the assembly and trim out excess soil from the two ends of the core with a spatula.

☆ Transfer the core into a container or remove the soil from the core into a labelled sample bag for transporting to the laboratory

☆ Soil samples are dried in a hot air oven at 105–110°C till constant mass is attained. The dried soil samples are weighed and their masses are recorded on the observation sheet.

☆ The diameter and length of the cores are also measured using bernier callipers for calculating the volume of the core. The bulk density is then calculated by the formula given above.

Note

In case of swell-shrink soils taking the fresh mass (core + moist soil) is essential as bulk density values change with change in soil water content and defining the soil water content at the time of sampling assumes significance.

3.3 Determination of Soil Reaction (pH)

Definition

pH is the logarithin to the base 10 of the reciprocal of the hydrogen ion (H^+) activity.

$$pH = \log_{10} 1/\ H^+$$

pH of a soil indicates its medium of reaction *i.e.* alkaline, neutral and acidic. Soil pH mainly governs the nutrients availability.

Method of Determination

The method of pH determination are classified into:

1. Colorimetric method (Kuhn's method).
2. Potentiometric method .

3.3.1 Colorimetric Method

This method is based on the assumption that if an indicator gives the same colour in two different solutions they will have the same pH value. The indicators are weak organic acids and bases. They give one colour in the acid soln. and another colour in the alkaline solution. This property of the changes in their colours with the change in pH of the medium makes them useful in the neutralization of acids and bases.

Principle

When a soil suspension is shaken vigorously with pure $BaSO_4$ the letter flocculates the soil colloids and leaves a clear and colourless solution. Then soil reaction is denoted by the colour of the indicator which is not adsorbed by the soil.

Procedure

☆ Place one cm thick layer of $BaSO_4$ in a test tube and add a layer of soil about 3 cm deep and fill with CO_2 free water to a depth of 9-10 cms.

☆ Add 0.5 ml of a suitable indicator and fill the tube to 15 cm depth with water., shake vigorously for about a half minute. Compare the colour of the suspension with colour on the lovibond comparator.

B.D.H. indicators are generally used to give approximate pH. For this purpose take a quarter spoonful of soil and little of $BaSO_4$ in flat dish, then add a few drops of the indicator. Tilt the dish to one side and observe the colour of the supernatant. It will indicate the approximate pH of the soil.

3.3.2 Potentiometric Method

Principle

The measurement of pH depends on the measurement of potassium develop across and class electrode on account of the differences in the activity of hydrogen ions in an out of the electrode.

Under these conditions when potentials of the two sources become equal, no current passes through the galvanometer. The needle is exactly brought to zero and the unknown potential of the cell containing the test solution is noted. It is equal to the potential of the external source. This potential (E) is defined by the nernst equation

$$E = \left(\frac{RT}{nF} \right) \log \left(\frac{K}{(M)} \right)$$

where,

R: Gas constant (8.31 joules)

T: Absolute temperature (298°K or 25°C)

N: Valancy of ion

F: Faraday constant (96,500 coulomb's)

M: Activity of ion to be measured

Apparatus

pH meter

Beakers

Glass rod

Pipette

Reagents

Buffer Solution: These are prepared by grinding of buffer tablets of pH 4, 7 and 9.2 dissolving separately in distilled water and make the volume to 100 ml using distilled water.

Procedure

> ☆ Take 10 gms of the soil and add 25 ml distilled water.

> ☆ Stir the suspension with rubber tipped glass rod at regular interval of 30 minutes dip the glass and colomel electrode in the soil suspension.

> ☆ Connect the electrodes to the pH meter which has already been calibrated with a standard buffer of known pH.

> ☆ Record the pH of the soil solution.

pH	Rating	Inference
<6.5	Acidic	Requires liming
6.5-7.5	Normal	No treatment needed
7.5-8.5	Saline/calcarious	Requires leaching of soluble salts
>8.5	Alkaline	Requires application gypsum

3.4. Determination of Electrical Conductivity

The Electrical conductivity (EC) or conductance is the reverse of resistance

$$EC = 1/R = 1/ohm$$

The resistance is defined by the Ohm's law as the ratio of electrical potential in volts and strength of the current in ampere.

$$E/C = R$$

Volts/Current = Resistance in Ohm's.

E.C. is expressed in reciprocal ohm's or mhos/cms. As the value of electrical conductance for soil solutions are very small, it is therefore convenient to express them in millimhos/cm. The values are often expressed calibrated at a temperature of 25°C.

Principle

A simple wheatstone bridge circuit is used to measure electrical conductivity by null method (Figure 3.2). In this two fixed resistances r_1 and r_2 and a variable resistance r_3 are connected in a branched circuit with conductance cell having resistance r_4. The variable resistance r_3 is adjusted until a minimum or 0 current flows through the AC galvanometer.

At equilibrium $r_1/r_2 = r_3/r_4$

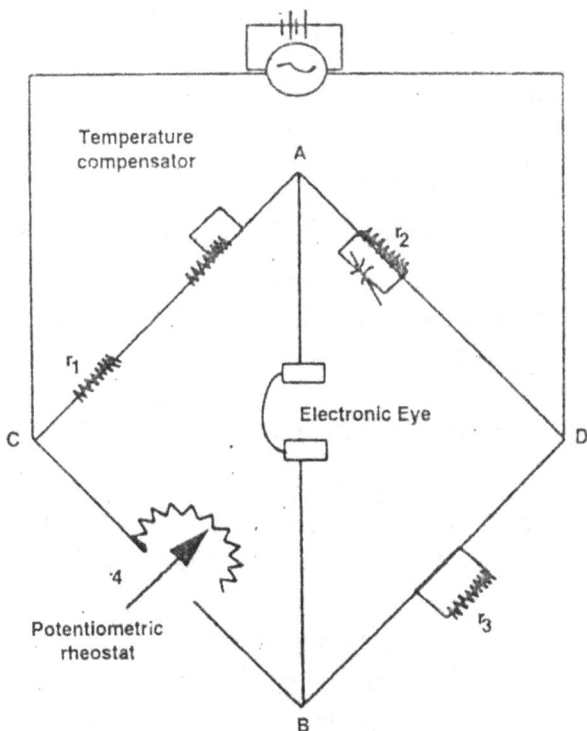

Figure 3.2: Wheatstone Bridge

$$r_3 = \frac{r_1 \times r_4}{r_2}$$

Apparatus

Conductivity meter

Beaker

Reagent

Standard Potassium Chloride Solution (0.01 M): Dissolve 0.756 g of dry KCl in one litre of distilled water. At 25°C it gives electrical conductivity of 1.413 ds/m

Procedure

☆ Take 10 g soil in a 100 ml beaker and add 20 ml distilled water in and shake for 2 hrs.

☆ Rinse the conductivity cell with distilled water and then with acetone and let it dry Dip the cell in the solution so that the electrodes are well immersed.

☆ Dip the cell in the solution.

☆ Balance the galvanometer or the magic eye of the conductivity meter and read the conductivity of the soil solution.

3.5 Determination of Cation Exchange Capacity

The capacity of negatively charged clays and organic matters to absorb cations by simple physical forces is called cation exchange capacity (CEC) of the soil The CEC of the soil not only depends upon the time and amount of soil colloids present, but is a function of soil pH as well. This is so because existence of the pH dependent negative charge on the surface of organic colloids and the type of clay mineral. The CEC of the soil is, usually express in c mol (P^+) kg^{-1}.

Method of Determination

The CEC is commonly determined by using distillation method.

Principle

The method most commonly used includes in that soil is first leached with neutral N Ammonium acetate solution to displace the exchangeable cations by NH_4^+ ions. The adsorbed NH_4^+ ions is finally determined by distillation the soil with MgO and the ammonia evolved during the distribution is absorbed by known volume of standard acid; the untreated acid molecules, being back titrated with standard alkali.

Apparatus

Erlenmeyer flask (500 ml).

Beakers 250 ml.

Distillation apparatus.

Hot place.

Porcelein dish.

Muffle furnace.

Burette.

Funnel.

Whatman's filter paper No. 42.

Balance.

Reagent

Neutral N-Ammonium Acetate Solution: Take 114 ml of glacial acetic acid (99.5 per cent) with distilled water to make approximately 1 litre. Then, add 138 ml of conc. NH_4OH, and add water to obtain a volume of about 1980 ml and adjust the pH 7.0 and dilute the solution to 2 litres.

Ethyl Alcohol (95 per cent): Dilute 250 ml of alcohol to 1000 ml, using distilled water.

Ammonium Chloride Crystals

Magnesium Oxide Power (MgO)

Sulphuric Acid (N/10)

Sodium Oxide (N/10)

Methyl Red Indicator: Dissolve 0.2 g of methyl red in 50 ml of alcohol and 50 ml of distilled water.

Procedure

- ☆ Weigh 10 g of soil sample in a flask.
- ☆ Add 50 ml NH_4OAc. Solution, mix thoroughly and keep it for half an hour with occasional stirring.
- ☆ Filter with Whatman No. 42. Transfer the soil completely on the filter paper.
- ☆ Continue to leach the soil with NH_4OAc solution.
- ☆ Keep the filtrate for the determination of total and individual exchangeable bases.
- ☆ To the soil on the filter paper, add a little amount NH_4Cl crystals and leach with alcohol.
- ☆ Continue washing of the residue till the filtrate runs free of chloride.
- ☆ Transfer the residue and filter paper to a 500 ml distillation flask. Add 0.5 g of MgO and 200-300 ml of distilled water. Connect the flask to the distillation unit.
- ☆ Place a 250 ml beaker containing 25 ml of H_2SO_4 N/10 under the condenser of the distillation apparatus to receive NH_3. Start distillation by

heating the distillation flask. Continue heating till no NH_3 is evolved.

☆ Back titrate the contents with N/10 NaOH.

Calculation

C.E.C. (me/100 g soil) = (B-T) x N x 1000 x 100/ W

where,

T: ml of N/10 NaOH used for sample

B: ml of N/10 NaOH used for blank

N: Normality of the acid

W: Weight of soil

3.6 Determination of Calcium Carbonate

3.6.1 Rapid Titration Method

The soil is treated with an excess of standard hydrochloride acid, destroying carbonates. The amount of excess acid is determined by titiration with standard sodium hydroxide solution using brom thymolblue indicator.

Apparatus

Burette

Pipette

Beaker

Reagents

Hydrochloric Acid (1 N)

Sodium Hydroxoide (1 N)

Thymolblue Indicator: Dissolve 0.1 g thymolblue indicator in 1.6 ml of 1 N NaOH and make the volume to distilled water

Procedure

- ☆ Weigh 5 g soil in a beaker and add 100 ml of 1 N HCl .

- ☆ Shake vigorously and then allow the soil to settle down.

- ☆ After 60 minutes take 20 ml of the supernatant in a beaker.

- ☆ Add 5 drops of thymolblue indicator and titrate it with 1 N NaOH.

- ☆ Also, carry a blank in the same manner (without soil).

Calculation

$$CaCO_3 \text{ (per cent)} = (B\text{-}A) \times 5$$

B: Blank reading.

A: Sample reading.

3.7 Determination of Lime Requirement of Soils

It indicates the amount of lime required, to bring the pH of an acidic soil to a desirable pH.

Liming Materials

Some of the liming materials used to raise the pH of acidic soils are:

Calcite or Calcium Carbonate $(CaCO_3)$

Dolomite $(CaCO_3. Mg CO_3)$

Calcium Oxide (CaO)

Hydrated Lime $[Ca(OH)_2]$

Lime Sulphur $(Ca S_5)$

3.7.1 Shoemaker *et al.* Method (1961)

Principle

The method involves equilibrating the soil with buffer solution at pH 7.5, whereby the reserve H^+ is brought into solution, which results in depression in pH of the buffer solution. A note of this is made and interpreted in terms of lime required to raise the pH to a desired value.

Apparatus

pH meter

Pipette

Beakers

Reagents

Buffer Solution: Dissolve 1.8 g nitrophenoal 3.0 g potassium chromate, 2.5 ml triethanolamine, 2.0 g. calcium acetate and 53.1 g $CaCl_2$ in 1 liter of distilled water and adjust the pH to 7.5.

Procedure

☆ Weigh 5 g soil in a beaker, add 12.5 ml distilled water and 10 ml buffer solution. Shake continuously for 10 minutes.

☆ For determining the pH, read the pH value of the buffer suspension by pH meter.

☆ Determine the pH of the soil sample. If the pH is below 6.3 then lime have to be recommended. Find the lime requirement from the Table 3.2.

3.8 Determination of Gypsum Requirement of Soil

Saline-alkali and alkali soils have a large amount of

sodium on their exchange sites. Gypsum ($CaSO_4 . 2H_2O$) generally is used to remove the exchangeable sodium from the soil complex. The amount of gypsum required to remove this exchangeable sodium from the soil complex is termed as gypsum requirement (GR).

Table 3.2: Lime Required to Bring the Soil to Indicated pH According to pH of the Soil Buffer Suspension

pH of Soil Buffer Suspension	Lime Required to Bring Soil to Indicated pH					
	pH 6.0		pH 6.4		pH 6.8	
	ton/acre	ton/ha	ton/acre	ton/ha	ton/acre	ton/ha
6.7	1.0	2.5	1.2	3	1.4	3.5
6.6	1.4	3.5	1.7	4.3	1.9	4.8
6.5	1.8	4.5	2.2	5.5	2.5	6.3
6.4	2.3	5.8	2.7	6.8	3.1	7.8
6.3	2.7	6.8	3.2	8.0	3.7	9.3
6.2	3.1	7.8	3.7	9.3	4.2	10.5
6.1	3.5	8.8	4.2	10.5	4.8	12.0
6.0	3.9	9.8	4.7	11.8	5.4	13.5
5.9	4.4	11	5.2	13.0	6.0	15.0
5.8	4.8	12	5.7	14.3	6.5	16.3
5.7	5.2	13	6.2	15.5	7.1	17.8
5.6	6.5	16.3	6.7	16.8	7.7	19.3
5.5	6.0	15	7.2	18.0	8.3	20.8
5.4	6.5	16.3	7.7	19.3	8.9	22.3
5.3	6.9	17.3	8.2	20.5	9.4	23.5
5.2	7.4	18.5	8.6	21.5	10.0	25.0
5.1	7.8	19.5	9.1	22.8	10.6	26.5
5.0	8.2	20.5	9.6	24.0	11.2	28.0
4.9	8.6	21.5	10.1	25.3	11.8	29.5
4.8	9.1	22.8	10.6	26.5	12.4	31.0

Principle

A know amount of calcium solution is equilibrated with soil and the amount of Ca left in solution is determined by titration with ethylene diamine tetra acetic acid (EDTA). The difference between amount of added Ca and Ca left in solution gives the amount of Ca exchanged.

Apparatus

Mechanical shaker

Burette

Pipette

Filter paper.

Conical flask.

Reagents

Saturated Calcium Solution: Shake 5 g $CaSO_4$ with 1 litre distilled water shake for 10 minutes and filter

Ammonium Chloride–Ammonium Hydroxide Buffer Solution: Dissolve 67.5 g of NH_4 Cl in 570 ml NH_4OH and makeup the volume to 1 litre with distilled water

Erichrome Black T Indicator: Dissolve 0.5 g of HCl and 4.5 g. of hydroxylamine–hydrochloride in 100 ml of 95 per cent of ethanol. It is called EBT indicator

Standard $CaCl_2$ Solution (0.01 N): Dissolve 0.5 g of $CaCO_3$ in 10 ml of dilute HCl and makeup to 1 liter with distilled water

Standard EDTA Solution (0.1 N): Dissolve 2 g of EDTA and 0.5 g of $MgCl_2.6H_2O$ in water and dilute to 1 liter. Stundardise the solution against the standard calcium solution

Procedure

    ☆ Weigh 5 g soil in a 250 ml conical flask and add 100 ml of the saturated $CaSO_4$ solution Shake for 5 minuntes and filter.

    ☆ Pipette 5 ml of the soil extract into a 100 ml conical flask and dilute to about 25 ml with distilled water.

    ☆ Add 0.5 ml of the NH_4OH buffer and 3-4 drops of EBT indicator and titrate with standard EDTA solution until the colour changes from wine red to blue.

    ☆ Similarly titrate 5 ml of the saturated Ca solutions as a blank.

Calculation

    Gypsum requirement (tons/acre) = 344 x N (A-B)

where,

    A: ml of EDTA for blank

    B: ml of EDTA for soil sample

    N: Normality of the EDTA

The factor 344 is derived as follows:

5 ml of the extract is taken for titration. But this has come form 100 ml. Thus, we get a factor 100/5. But this 100 ml has been used for extracting 5 gm. Soil, so we get another factor of 100/5. Therefore, total dilution becomes 100/5 x 100/5 and when we multiply by me, N (A-B), we get me/100 g. To convert me/100 g 408/acre multiply me/100/g soil by equivalent (86) x 20. When we change this to tons/acre thus:

    100/5 x 100/5 x 86 x 20/2000 = 344 ton/acre.

Chapter 4
Plant Analysis

4.1 Analysis of Plant Tissue

Leaf is a principle site of plant metabolist, therefore, changes in nutrient supply are reflected in the composition of plant tissues. These changes more pronounced at certain stage of development and the leaf nutrient concentration at specific growth stages are related to the crop performance.

The plant analysis has been used as a diagnostic tool to determine the nutrient causes of plant disorders/diseases. The plant analysis constitutes (*i*) the collection of the representative plants parts at the specific growth stage (*ii*) washing, drying and grinding of the plant tissue, (*iii*) oxidation of the powered samples to solubilize the element or elements, (*iv*) estimation of different elements and (*v*) interpretation of the status of nutrients with respect to deficiency/efficiency/toxicity on the basis of known critical concentrations. The plant materials can be oxidized either by dry ashing at a controlled high temperature in a muffle furnace or by wet digestion in an acid or a mixture of two or more acids.

4.2 Nitrogen

Nitrogen is one of the major constituents required for nutrition of plants. It plays an important role in synthesis of proteins, and is involved in various metabolic activities in plants and thus for is required development of plant/crop. N content of plants varies from 0.2 per cent to 6 per cent of dry weight basis.

Tissue tests helps to diagnose N deficiency/ toxicity in plants.

Total N content of plant tissues is, generally, determined by the 'Kjeldahl method'.

Principle

The method of determination, essentially, involves 3 successive phases:

(*a*) Digestion of organic material to convert N into NH_3.

(*b*) Distillation of the digested materials.

(*c*) Titration of the distilled product.

Digestion of the organic material is accomplished by boiling the sample with conc. H_2SO_4, in the presence of catalysts which accelerate the rate of digestion, and salts (K_2SO_4 or Na_2SO_4), which raise the digestion temperature. The organic material decomposes into several components: $C \rightarrow CO_2$, $H \rightarrow H_2O$, and $N \rightarrow NH_3$.

N is transformed into NH_3, during digestion and because of the excess H_2SO_4, it remains in the form of ammonium sulphate.

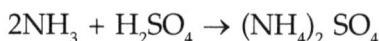

$$2NH_3 + H_2SO_4 \rightarrow (NH_4)_2 SO_4$$

NH_3 is determined by distillation, The NH_3 gas, so liberated, is absorbed in a known amount of boric acid, the quantity is being determined by back-titration with a standard acid.

Titration $\qquad$ $NH_3 + H_2O \rightarrow NH_4 OH$

$$4H_3BO_3 + NH_4OH \rightarrow (NH_4)_2B_4O_7 + 7H_2O$$

Ammonium tetraborate

$$(NH_4)_2 B_4O_7 + H_2SO_4 + 5H_2O \rightarrow 4H_3BO_3 + (NH_4)_2 SO_4$$

Apparatus

Kjeldahl digestion assembly

Kjeldahl flask

Distillation assembly

Conical flask or beaker

Pipette

Reagents

Concentrated Sulphuric Acid and (0.1 N) H_2SO_4

Granulated Zinc (Zn)

Digestion Mixture: 10 g of $CuSO_4$ $5H_2O$ (Oven dried), 3 g HgO and 1 g Se powder

Potassium Sulphate

Sodium Hydroxide (30 per cent)

Mix Indicator: 0.1 g bromocresol green and 0.07 g methyl red in 100 ml of ethanol

Boric Acid (2 per cent): dissolve 20 g boric acid, (H_3BO_3) in about 900 ml of hot water. Cool and add 20 ml of a mixed indicator solution. Add 0.1 N NaOH solution dropwise, until the colour changes to reddish and dilute to 1 litre with distilled water.

Procedure

(a) Digestion

☆ Weigh 1 g of grinded plant sample in a Kjeldahl flask.

☆ Add 25 ml of concentrated H_2SO_4, 20 ml of distilled water, 1 g of digestion mixture and 10 g of K_2SO_4.

☆ Place the flask in the digestion unit. Digest the content of the flask at low heat to prevent frothing.

☆ After 15-20 minutes, gradually raise the heat until the content becomes clear and coloured pale green or blue.

(b) Distillation

☆ Cool the content, add about 200 ml of distilled water and swirl the flask for about 2 minutes, and take the supernatant liquid into a distillation flask.

☆ Add about 50 ml of water to the digestion flask, and take the water extract into the distillation flask. Repeat this process for at least, 4 times.

☆ Add about 150 ml of 30 per cent of NaOH, slowly along the side of the distillation flask. Add 2 pieces of Zn to prevent superheating of the soil extract.

★ Distill NH_3 into a 25 ml of boric acid-indicator solution contained in a conical flask, and place the flask below the condenser, so that the discharging tube of the condenser is immersed in the boric acid solution.

★ When no more NH_3 is received (test with a wet red limits paper–not turning blue), stop the distillation and proceed for titration.

(c) Titration

★ Ammonium tetraborate is formed during the distillation, which is back titrated with 0.01 N H_2SO_4, releasing boric acid with the formation of $(NH_4)_2$ SO_4. The disappearance of blue colour indicates the endpoint of the titration. Run a blank (without plant material) to avoid contamination and to ensure precision.

Observations and Calculation

★ Weight of the plant material = 1 gm
★ Normality of H_2SO_4 = 0.1
★ Volume of H_2SO_4 used for sample = S ml
★ Volume of H_2SO_4 used for blank = B ml
★ ml of H_2SO_4 used for sample = S–B
★ 1 ml of 1 N H_2SO_4 = 0.014 g N
★ 1 ml of 0.1 N H_2SO_4 = 0.0014 g N
★ N in sample (per cent) = (S–B) x 0.0014 x 100

**Average Range of Variation of
Different Nutrient Elements in Plants**

Elemental	% of Dried Substance	Element	ppm in Dried Substance
N	0.2-6	Co	0.01-5
P	0.05-2	Mo	0.05-20
K	0.10-12	Cu	1.00-50
Ca	0.10-4	B	1.00-100
Mg	0.01-2	Zn	1.00-200
S	0.05-1.5	Fe	10.00-250
Cl	0.04-2	Mn	10.00-1000

4.3 Dry Ashing

Principle

In this method, the vegetative material is destroyed under the influence of temperature from 300 to 450°C initially, and until the material is well dried; this is followed by a gradual increase in temperature to 550°C, till a grayish-white ash is formed. The ash is finally dissolved in an acid, for subsequent analysis of mineral constituents.

Water, N, C, H, O, S etc. lost at this temperature, in case of P, magnesium acetate ignition method is followed to check its volatilization loss, which, otherwise, is prominent in simple combustion of plant material.

Apparatus

Porcelain dish or silica dish

Muffle furnace

Hot plate

Volumetric flask ashless

Whatman No. 42 filter paper

Reagents

Hydrochloric Acid (6 N)

Magnesium Acetate Solution (0.5 N)

Procedure

☆ Weigh 1 g of grinded plant material in a silica or porcelain dish.

☆ Place it in a muffle furnace at 300 to 450°C for about 15-20 minutes, so that the material is well-dried. Gradually, raise the temperature to 550°C and equilibrate till a greyish-white ash is obtained.

☆ In case of P, its volatilization loss is precluded by mixing the plant tissue with 5 ml of 0.5 N magnesium acetate solution.

☆ Cool, and dissolve the ash in 10 ml of 6 N HCl.

☆ Evaporate the solution to dryness on a hot plate.

☆ When the solution is dry, add 5 ml of 6 N HCI and 20 ml of water; warm to dissolve the soluble constituents and filter through an acid-washed filter paper into a 100 ml volumetric flask.

☆ Dilute to volume, and mix thoroughly.

☆ Determine Ca and Mg by the EDTA or versenate titration method; K by ammonium acetate method and P by vanadomolybdate method.

Precautions

A sudden heating is not advocated, as it leads to loss through volatilization, formation of silicates complex and melting of the ash. The operation should, therefore, consist of two stages, the first stage is when the temperature does not exceed 450°C and the second stage is at 550°C until a white ash is obtained.

4.4 Wet Ashing

Principle

The methods consists in the oxidative destruction of the plant material, involving different types of digestion mixture; the commonest being nitric and perchloric acid digestion is continued until the acid liquid has been completely volatilized, and a colourless solution formed. The filtered solution is then used for various elemental analyses, using appropriate techniques.

Apparatus

Conical flask

Hot plate

Whatman No. 42 filter papers

Volumetric flask

Reagents

Concentrate Nitric Acid: perchloric acid (3:1)

Procedure

☆ Place 1 g of the ground plant material (20-mesh) in a 150 ml conical flask.

☆ Add 10 ml of concentrated diacid mixture and place the flask on a hot plate at 190°C; continue the pre-digestion until the solution has nearly evaporated. Pre-digestion is required in most cases to avoid bumping or violent reaction, on addition of di-acid mixture to the plant material.

☆ Remove the flask, cool and add 5 ml of the diacid mixture; replace the flask on the hot plate, and continue digestion till a colourless solution is obtained. In case the liquid turns brown, add another 5 ml of the acid mixture, and digest.

☆ When the liquid is completely clear, reduce the volume of the acid by heating until flask contains only moist residue.

☆ Cool the flask and add 25 ml of distilled water.

☆ Shake the solution into a 50 ml volumetric flask, and dilute to the mark with distilled water; the aliquot is being used for pertinent nutrient analysis using standard procedure

4.5 Determination of Phosphorus

4.5. 1 Vanadomolybdate Method

Principle

A suitable portion of the plant material is transferred into a volumetric flask, and allowed to react with vanadomlybdate in an acidfield solution. A characteristic yellow chromogen of the vanadomolybdophosphoric acid is formed. The intensity of colour is measured colorimetrically.

Apparatus

Photoelectric colorimeter

Volumetric flask

Pipette

Reagents

1. *Vanadomolybdate Reagent*: It is prepared by mixing the following two solutions:

Solution A: dissolve 25 g of ammonium molybdate, $(NH_4)_6MO_7O_{24} \cdot 4H_2O$ in 400 ml in warm distilled water and cool.

Solution B: Add slowly 1.25 g of ammonium metavandate to 300 ml of boiling water, and cool, thereafter add to it 250 ml concentration HNO_3, and cool. Pour solution B into a 1 litre volumetric flask, and add solution A. Mix well, and dilute to volume of 1 liter with distilled water.

2. *Phosphate Standard Solution*: Dissolve 0.2195 g of KH_2PO_4 and dilute to 1 litre. This solution contains 50 ppm P. Use this stock solution for the preparation of working standards of P.

Procedure

☆ Take 5 ml of the ash solution in a 50 ml volumetric flask.

☆ Add 10 ml of vanadomolybdate and make the volume with distilled water.

☆ Read the intensity of yellow colour at 450 nm using blue filter.

Preparation of the Standard Curve

Pipette 0, 1, 2, 4, 6, 8, 10 ml of 50 ppm solution separately in 50 ml volumetric flasks. Add 10 ml of vanadomolybdate reagent, and proceed for developing the colour in the same way as described above.

Observations and Calculation

☆. Weight of the plant material   = 1 g

☆ Volume of plant digest   = V ml

☆ Volume of aliquot taken = V_1 ml

☆ Final volume made = 50 ml

Phosphorus

☆ Transmittance (per cent) as read from the colorimeter = T (say)

☆ P (ppm) as read from the standard curve =Y (say)

☆ P (ppm) in the plant material = Y x 50/1 x V/V_1

4.6 Determination of Potassium

Principle

Oven dry plant sample grinded and past through 40 mesh sieve and digested in diacid or triacid or by dry ashing. Digested plant sample is diluted and K in the plant sample is injected into the flame photometer. The potassium atom excited when passing through a flame to the higher orbit. Such atoms release energy of a wavelength and give spectral lines for that element and is proportional to the concentration of the element.

Reagents

Standard Stock Solution (1000 ppm K): Dissolve 1.906 g of AR grade KCl in distill water and make the volume to 1 liter.

Working Standard K Solution: To prepare 0, 5, 10, 15 and 20 ppm K solution by taking 0, 2.5, 5.0, 7.5 and 10 ml of 1000 ppm K solution in 500 ml of volumetric flask and make the volume to the mark with distilled water.

Procedure

☆ Digest the plant sample in diacid (HNO_3-$HClO_4$) till it is converted into white precipitate.

☆ Calibrate the instruments. Atomise the intermediate K standard and note down the corresponding flame photometer reading.

☆ Prepare the standard curve by plotting the flame photometer reading corresponding to the K concentration.

☆ Determined the K content in the plant sample

☆ Similarly run the blank.

Observations and Calculation

☆ Weight of plant sample = 1

☆ Total volume of plant digest = 50

☆ Flame photometer reading = F (say)

☆ K (ppm) as read from the standard curve = U (say)

☆ K (ppm) in the given plant material = F x U x 50

$$\text{☆ K (per cent)} = \frac{F \times U \times 50}{10000}$$

4.7 Determination of Calcium and Magnesium

Calcium in the plant sample is determined by titration with 0.01 N EDTA in the presence of ammonium purpurate. Whereas, Calcium and magnesium can be determined in digest by titration with EDTA using Erichrome Black T indicator (EBT).

Reagents

Calcium Chloride Solution (0.1 N): Dissolve 0.4 g of pure calcium carbonate in 10 ml of (1+3) HCl and dilute to 1 litre with distilled water.

EDTA (0.01 N): Dissolve 2 g of EDTA and 0.05 g $MgCl_2$ $6H_2O$ in one liter of distilled water. Standardize the solution against 0.01 N CaCl$_2$ using ammonium purpurate indicator.

Sodium Hydroxide (4 N): Dissolve 16 g of sodium hydroxide in 1 liter of distilled water.

Potassium Cyanide Solution (1 per cent): Dissolve 1 g of KCN in 100 ml of distilled water.

Ammonium purpurate mix 0.5 g ammonium purpurate with 100 g potassium sulphate power.

Procedure

- ☆ Take 10 ml of plant extract in 150 ml Erlenmeyer flask. Evaporate to dryness.
- ☆ Cool it and add 25 ml distilled water to dissolve the residue
- ☆ Add pinch of disodium diethyl dithio carbamate crystals and 5 drops of 1 per cent KCN solution.
- ☆ Add 4 N NaOH drop by drop to raise the pH 12.
- ☆ Add pinch of ammonium purpurate indicate
- ☆ Titrate against EDTA till the colour changes from pink to purple.

Observations and Calculation

Calcium

- ☆ Volume of aliquot taken for analysis = V_1 ml
- ☆ Normality (N) of EDTA used = 0.01
- ☆ Volume of EDTA used in titration (titre value) = V_2 ml
- ☆ Meq. of Ca/100 of plant material = $(0.01 \times V_2) \times V_1 \times 100/1$

Determination of Calcium + Magnesium

Magnesium formed a complex with EDTA in the presence of EBT at the end point colour changes from vine red to blue.

Reagents

Magnesium Chloride (0.01 N): Dissolve 0.5 g of $MgCl_2 \bullet 6\ H_2O$ in 500 ml of distilled water.

Buffer Solution: Dissolve 67.5 g of ammonium chloride in 570 ml of ammonium hydroxide solution and make the volume to 1 liter with distilled water.

EBT Indicator: Take 0.5 EBT and 45 g hydroxylamine hydrochloride in 100 ml of 95 per cent ethyle alcohol.

Hydroxylamine Hydrochloride Solution 5 per cent: Dissolve 5 g of hydroxylamine in 100 ml of distilled water.

Potassium Ferocyanide Solution 4 per cent: Dissolve 4 g of potassium ferrocyanide in 100 ml of distilled water.

Triethanolamine (50 per cent): Transfer 50 ml of TEA in 100 ml of distilled water.

Calcium Chloride Solution: Same as discuss in calcium determination.

EDTA (0.01 N): Same as discuss in calcium determination.

Procedure

☆ Take 10 ml of aliquot in 150 ml Erlenmeyer flask evaporate to dryness.

☆ Add 25 ml distill water to resolve the residue.

☆ Add 1 drop of 0.01 N $MgCl_2$ and 5 drops each, of KCN 1 per cent, potassium ferocyanite (4 per cent), 5 per cent hydroxylamine hydrochloride.

☆ Add 2 ml of buffer solution, 2 drops of EBT

☆ Titrate against EDTA solution

☆ Similarly run the blank.

Calcium + Magnesium

☆ Volume of aliquot taken for analysis = V_1 ml

☆ Volume of EDTA used in titration (titre value) = V_3 ml

☆ Meq of (Ca + Mg/100 gm of plant material = (0.01 x V_3 x V_1 x 100/1

☆ Meq. of Mg/100 g of plant material = meq. of (Ca + Mg)/100 g of plant material – meq. of Ca/100 g of plant material

4.8 Determination of Micronutrient Cations (Zn, Mn, Cu and Fe)

Zinc, manganese, copper and iron are estimated in plant digest by dry ashing or from wet digestion by HNO_3 and $HClO_4$.

Principle

The atoms of metallic elements (Fe, Mn, Cu, Zn) absorb energy when subjected to radiations of specific wavelength. The absorption of radiation is proportional to the concentration of the element.

Apparatus

Atomic absoption spectrophotometer

Analytical balance

Measuring cylinder

Hot plate

Conical flasks

Polyethylene/glass funnels

Polyethylene bottles

Measuring flaks.

Pipettes graduated

Whatman No. 1 filter paper

Reagents

Concentrated Nitric Acid (HNO_3)

Concentrated Perchloric Acid ($HClO_4$)

Standard Solutions: The standard solution of different micronutrients cations should be prepared as detailed under soil analysis.

Working Standard Solutions: They should be prepared as described under soil analysis.

Nitric Acid (0.25 N): Dilute 16 ml of concentrated HNO_3 to 1 liter with distilled water.

Procedure

Digestion of Plant Material

Plant material is digested either in HNO_3-$HClO_4$ mixture or ashed dissolved in acid. Also carry a blank digestion using all steps (without plant material).

Analysis

☆ Zn, Cu, Fe and Mn concentration in plant digest can be determined by atomic absorption spectrophotometer as follows:

☆ Set zero of the instrument with blank (blank digest).

☆ Feed standards of element to be determined to the AAS to standardize the instrument for element in sample

☆ Feed plant digest and record the absorbance/concentration of the element.

☆ Repeat the above steps for each element.

Calculations

$$\text{Zn, Cu, Mn or Fe in plant sample} = \frac{\text{AAS reading (mg } l^{-1}) \times \text{dilution}}{\text{Weight of plant tissue (g)}}$$

4.9 Determination of Boron

Total boron (B) in plant material is determined by ashing of plant sample followed by the dissolution of the ashed material and B is analysed by spectrophotometer.

Apparatus

Spectrophotometer

Analytical balance

Muffle furnace

Porcelain dish

Pipette

Watch glass

Volumetric flasks

Conical flasks

Filter paper

Whatman No. 1

Water bath

Reagents

Calcium Oxide (CaO), powdered

Buffer Solution: Dissolve 250 g of ammonium acetate and 15 g of disodium salt of EDTA (ethylene diamine tetraacetic acid disodium salt) in 400 ml of distilled water and slowly add 125 ml of glacial acetic acid and mix.

Azomethine-H Solution: Dissolve 0.45 g of Azomethine-H in 100 ml of 1 per cent L-ascorbic acid solution.

Standard B Solution (100 µg B ml⁻¹): Dissolve 5.716 g of boric acid (H_3BO_3) in a small volume of distilled water and make the volume to 1000 ml with distilled water.

Working Standard Solution of B (20 µgB ml⁻¹): Dilute the stock solution by 50 times (take 2 ml and make it to 100 ml with deionized water and mix).

Procedure

(a) Ashing and Extraction of the Sample

☆ Weigh 0.5 g of grinded plant sample

☆ Add 0.1 g calcium oxide powder and mix and transfer into the porcelain crucible.

☆ Place the crucible in a muffle furnace. Raise the temperature of the furnace slowly to a maximum of 550 °C.

☆ Cool the content and added 3 ml of dilute HCl (1 + 1)

☆ Heat on a water bath for 20 minutes

☆ Transfer the contents of the crucible to 25 ml volumetric flask, make the volume to the mark with distilled water and filter through a filter paper.

(b) Estimation of B in the Plant Digest

☆ Take 1 ml of aliquot into a polyethylene tube or small beaker.

☆ Add 2 ml of buffer solution and mix the contents thoroughly

☆ Add 2 ml of azomethine–H reagent and again mix the contents thoroughly.

☆ Let the mixture stand for 30 minutes and then measure the colour intensity at 420 nm by spectrophotometer

(c) Preparation of Standard Curve

☆ Take 0, 10, 20, 30, 40 and 50 ml of the working standard solution of B into 100 ml volumetric flasks and make the volume upto the mark with distilled water in order to obtain 0, 2, 4, 6, 8 and 10 µg B/ml respectively.

☆ Take 1 ml each of the final standard solutions into a separate small beaker.

☆ Then follow steps 2 to 4 mentioned under estimation of boron in the plant digest.

Observations and Calculation

☆ Weight of plant sample taken for ashing = 0.5 g

☆ Final volume made after extraction of B in acid = 25 ml

☆ Dilution = 50 times

☆ Absorbance reading of the spectrophotometer = A

☆ Concentration of B (ppm) as read from the standard curve against A = C.

☆ Content of B (ppm) in the plant sample = C x 50.

4.10 Determination of Molybdenum

For the determination of total Mo in plants, sample is first digested with HNO_3 and $HClO_4$ as described by Purvis and Peterson (1954) and Mo in the digest is estimated by colorimetric method of Johnson and Arkley (1954).

Principle

Molybdenum forms an amber orange coloured complex when react with thiocyanate. The concentration is determined by colorimeter.

Apparatus

Analytical balance

Conical flasks

Hot plate

Pipettes

Volumetric flasks

Funnels

Filter paper, Whatman No. 1

Reagent

Concentrated HNO_3 (AR grade)

Perchloric Acid

Hydrogen Perioxide (H_2O_2)

Isopropyl Ether: The ether is washed before use with a wash solution comprising a mixture of one-third stannous chloride ($SnCl_2$), one third potassium thiocyaniate (KSCN) and one third deionized water. For washing, place reagent grade isopropyl ether in a separatory funnel and add wash solution equal to one tenth of the volume of ether taken.

Shake thoroughly and let organic phase separate from aqueous phase. Drain and discard the aqueous phase. Wash with 2 N HCl again taking one tenth of the volume of ether. Shake and discard the aqueous phase. Repeat this process four to five times.

Hydrochloric Ferric Chloride: Dissolve 0.5 g of hexahydrate ferric chloride ($FeCl_3 \cdot 6H_2O$) to 560 ml of concentrated HCl and dilute to one liter with distilled water. This solution is required for preparing standards and reagent blank.

Potassium Thiocyanate (40 per cent w/v): Suspend 40 g of stannous chloride dehydrate ($SnCl_2 \cdot 2H_2O$) in 20 ml of 6.5 N HCl. Add distilled water to dissolve it and make upto 100 ml. with distilled water.

Stannous Chloride (40 per cent): Dissolve 40 g of stannous chloride in 20 ml of 6.5 N HCl and make the volume to 100 ml with distilled water filter if turbid.

Standard Solution of Mo (100 ppm Mo): dissolve 75 mg of pure molybdenum trioxide (MoO_3) in 5 ml of 0.1 N sodium hydroxide and dilute to approximately 400 ml with distilled water. Make slightly acidic with HCl and make the volume to 500 ml with distilled water.

Working Standard Solution of Mo (1 µg Mo/ml): Dilute 10 ml of the stock standard solution of Mo to one litre to obtain 1 µ g Mo/ml.

Citric Acid (50 per cent): Dissolve 50 g of citric acid crystals in distilled water and dilute to 100 ml with distilled water.

Procedure

(a) Digestion of Plant Samples

☆ Weigh 1-2 g plant sample into a 150 ml conical flasks.

☆ Add 15 ml di-acid mixture HNO_3 and $HClO_4$ (4: 1) and place a funnel at the mouth of the flask. Preferably keep the contents of the flask overnight.

☆ Digest at low heat and raise the temperature slowly till white fumes of $HClO_4$ start appearing. Evaporate the contents to dryness.

☆ Cool and add 5 ml of HNO_3-$HClO_4$ mixture and again evaporate to dryness.

☆ Then add 1 ml H_2O_2 and allow the contents to dry and cool.

☆ Add 50 ml of distilled water, boil for one minute and add 10 ml concentrated HCl.

☆ Make the volume to 100 ml with distilled.

☆ Filter the solution and analyse for Mo.

(b) Analysis of the Digest

☆ Take 50 ml of the digest to a separatory funnel.

☆ Add 2 ml of 50 per cent citric acid solution and dilute the contents to 60 ml with distilled water.

☆ Add 2 ml of di-isopropyl ether and shake for 2 minutes releasing pressure as necessary.

☆ Discard the organic phase.

☆ Add 1.5 ml of 40 per cent potassium thiocynate solution and 1.5 ml of stannous chloride solution.

☆ Discard the aqueous phase and read the intensity of colour with a colorimeter at 470 nm.

(c) Preparation of Standard Curve

☆ Take 0, 0.5, 1, 2, 4 and 6 ml of solution containing 1 ppm Mo in separatory funnels.

☆ Add 10 ml of 6.5 N HCI-Fe Cl_3 solution to each separatory funnel.

☆ Dilute the HCl-molybdate solution to 45 ml with distilled water.

☆ Add 2 ml of 50 per cent citric acid solution to 45 ml with distilled water.

☆ Add 2 ml of 50 per cent citric acid solution and dilute the content to 60 ml with distilled water.

☆ Add 2 ml of di-isopropyl ether and shake vigorously for 2 minutes, releasing pressure as necessary.

☆ Allow 10 to 15 minutes for phase separation and discard the organic phase.

☆ Add 1.5 ml of 40 per cent KSCN solution and then 1.5 ml of stannous chloride solution and mix again.

☆ Add 10 ml of di-isopropyl ether and shake well.

☆ Measure the colour intensity at 470 nm.

Observations and Calculation

☆ Weight of plant sample taken for digestion = 1.0 g

☆ Final volume made after digestion = 100 ml

☆ First dilution = 100 times

☆ Volume of the filtrate taken in the separatory funnel = 50 ml

☆ Volume of the organic solvent used for extraction = 5 ml

☆ Second dilution = 0.1 times

☆ Total dilution = 10 times

☆ Absorbance read by spectrophotometer = A

☆ Concentration of Mo (ppm) observed from the standard curve against A = C

☆ Amount of Mo (ppm) in the plant sample = C x 10

4.11 Interpretation of the Plant Analysis of Data

The plant analysis data are interpreted using established concentration ranges for deficiency, low, sufficiency, high and excess (toxic) categories; standard values; critical values; ratios, distribution and difference approach.

Element	Content in % Dried Substance		
	Deficiencies	Optimum	Toxicity
P	<0.1	0.3-0.6	>1.0
K	<1.2	2-4	>5
Ca	0.15	1-4	-
Mg	<0.1	0.2-0.8	>2

Table 4.1: Micronutrient Critical Levels for Crop Plants

Crop	Nutrient (ppm)			
	Zn	Cu	Fe	Mn
Wheat	10	3	10	10
Rice	10	3	70	20
Maize	11	3	10	15
Cotton	20	4	30	30
Sugarcane	15	1	5	15
Gram	20	5	50	64

Contd...

Table 4.1–Contd...

Crop	Nutrient (ppm)			
	Zn	*Cu*	*Fe*	*Mn*
Groundnut	20	6	50	50
Sunflower	20	4	60	41
Berseem	15	5	30	15
Potato	15	2	11	10
Tomato	25	5	100	50
Cauliflower	15	3	50	25
Peas	40	10	50	40
Kinnow	15	4	40	20
Grapes	20	5	35	30
Guava	20	7	45	21
Peach	15	3	60	20
Pear	10	5	60	20
Litchi	15	5	50	50
Mango	20	10	70	60

Chapter 5

Advance Methods of Soil and Plant Analysis

5.1 Plasma Atomic Emission Spectrophotometer (ICAP-AES)

Atomic emission spectroscopy, employing inductively coupled plasma (ICP) as the source of energy of major, and trace elements in virtually all types of soil and plant materials with excellent sensitivity and emission stability. In early 1960's Reed found the plasma to have (*i*) high temperatures (*ii*) capability of being sustained in noble gas environments and (*iii*) freedom from contamination from electrodes which are not required. Since, then a lot of research has been conducted on the use of ICP (Fassel, 1978) and at present it has been found to be the best technique for elemental analysis.

An ICP is a high energy optically thin excitation source. Power from a radio frequency generator is coupled to a flow of ionized argon gas inside a quartz tube encircled by an induction coil. To initiate the plasma argon is ionized by a momentary high voltage discharge. The ionized gas passing through the high frequency magnetic field absorbs energy. This causes further heating and ionization to form a ball of electrically conducting gas of plasma. Liquid samples in the form of an aerosol are injected into the high temperature environment of plasma. Here the analyte forms free atoms which emit spectra. More than 70 of the elements in periodic table are capable of being determined by ICP.

Plasma is a major excitation mechanism which is produced electrically in a carrier inert gas such as nitrogen or argon. The energy levels of argon make this element especially well suited for plasma production and this gas has the added advantage of chemical inertness.

A PLASMA can be defined as a neutral gas containing significant numbers of both positive and negative ions or free electrons a plasma can only be created and maintained by the continued injection to ensure that new ions are created fast enough to compensate for those that are continually recombining to form neutral atoms.

Two types of argon plasma are in use. One energized by radio frequency are coupled to the gas through electromagnetic induction is called inductively coupled plasma, whereas the other used d-c excitation (DCP). ICP is more of the two types but d-c plasma are considerably less expensive. The temperature of DCP plasma may reach as high as 5000 °K.

The ICP source consists of three quartz tubes through which argon flows and a two-or-three-turn gola plated copper coil surrounding the tubes near the upper end.

Alternating current at a frequency of 27.14 Megahertz and power produced by means of an auxiliary spark circuit a heavy current is caused to flow in a circular path in the ionized gas powered plasma by magnetic induction. This raises the temperature of the resulting plasma to 10,000°K. This is above the softening point of quartz. The plasma torch is protected from destroying itself by using the flow of argon itself as coolant. The bulk of the argon enters the outer tube at a tangential angle, so that it swirls through the annular space at high speed thus moderating the temperature. The hot plasma tends to stabilize in the form of a toroid serves to prevent overheating.

The sample is aspirated in a nebulizer and is carried by a slower stream of argon directed centrally towards the "hole in the doughnut". Here the sample is heated by conduction and radiation and may reach 7000°K where it is completely atomized and excited. Loss of analyte atoms by ionization a source of difficulty in flame and spark does not occur significantly in ICP spectroscopy, presumably because of the presence of the more easily ionized argon atoms.

Spectrophotometers using plasma as the source of energy are inherently different from those using spark or arc. Some models operate in sequential mode wherein the wavelengths for all desired elements are scanned in order. Other act in a simultaneous mode in which many elements are detected at the same time with multiple phototubes.

Advantages of ICP-AES

1. High quality multi-element simultaneous analysis.
2. Refractory elements like P, B, W, Zr and U can be determined.

3. High linear dynamic range.
4. Exceptional stability over long period of operation.
5. Excellent detection limits–0.1 to 10 ppb.
6. Reduced inter element effects.
7. Background emission interference is nil.
8. Self absorption of radiation is nil.
9. Loss of analyte atoms by ionization does not occur significantly because of the presence of more easily ionized argon atoms.

Preparation of Soil Samples for on ICP-AES

(*i*) Soil Analysis

Total elemental analysis

1. Sodium carbonate digestion (Na_2CO_3)
2. Perchloric acid digestion ($HClO_4$)
3. Hydrogen floride digestion (HF)

Mix the soil (100 mesh) with anhydrous sodium carbonate (1:3 to 1:10) thoroughly and transfer to platinum crucible. Dissolve in 100 ml of 1N HCl. Take to dryness to dehydrate the silica. Add 20 ml of 3 N HCl and heat to dissolve the contents. Filter and make volume to the mark. The extract can be used for the determinations of total P, Ca, Mg, Zn, Cu, Fe, Al, Ti (Chapman and Pratt, 1961).

Where total phosphorus is the only constituent to be determined in soils it can be completely extracted by digestion with 72 per cent perchloric acid in the ratio of 1: 2. Two g soil is taken with 4 ml of perchloric acid in 50 ml beakers covered with watch glass and put on a hot plate. The digestion is continued till the soil changes its colour to white. It should not be taken to dryness. It is now cooled and diluted to desired volume before filtration.

For some elements like calcium and manganese the digestion of soil in HF is preferred. One g of soil taken in Pt crucible is treated with 2 ml of water and 1 ml of perchloric acid before 5 ml of HF is added. The contents are evaporated to dryness. The residue is taken up in 1 ml of HCl and 5 ml of water and heated to dissolve. After cooling the desired volume is made and filtered.

For Mn 0.5 g soil is treated with 4 ml HF and 2 ml of sulphuric acid. The digestion is conducted at low temperature till the soil colour changes to white.

The HF extracts can be analysed on ICAP-AES only if the sample is evaporated to complete dryness after digestion so that there is no free HF which would otherwise eat away the glass nebulizer as well as the quartz torch tubes. Some ICP-AES instruments have the Teflon nebulizer which is resistant to HF.

(*ii*) Available Elemental Analysis

Ammonium Bi-cabonate-DTPA (AB-DTPA) reagent (Soltanpour and Schwab, 1977).

For simultaneous multi-element determination, obviously single element extraction solutions are not useful. Soltanpour and Schwab (1977) developed an 1 M NH_4HCO_3 (0.005 M) DTPA (AD-DTPA) solution (pH=7.6) for simultaneous extraction of P, K, Zn, Cu, Fe and Mn from soils. This test was modified by Soltanpour and Workman (1979) to omit carbon block, which sometimes contaminate the sample and adsorb metal chelates. After extraction ICP-AES is used to simultaneously analyte these extracts for the above mentioned elements. Experience has shown that AB-DTPA solution should be acidified to get rid of the carbonate-bicarbonate matrix in order to prevent clogging of the capillary tip (Soltanpour *et al.*, 1979; Arora and Hundal, 1995).

Theoretical Basis

DTPA, chelates the metals, and ammonium ions exchanges with potassium and bring it into solution. During shaking the original pH (7.6) rises due to evolution of CO_2. As the pH rises bicarbonate changes to carbonate. The carbonate ions precipitate calcium from calcium phosphates and thus increases to solubility of phosphorus. This method has been found to be highly correlated with ammonium acetate method for potassium, sodium, bicarbonate method for phosphorus and DTPA method for Zn, Cu, Fe and Mn. The ranges and sensitivities are same as those for DTPA, sodium bicarbonate and ammonium acetate methods for micro-nutrient potassium and phosphorus, respectively.

Extracting Solution

Add 1.87 g of diethylene triamine penta acetic acid (DTPA) to 800 ml of distilled water taken in 1 liter corning beaker. Add approximately 2 ml of 1: 1 NH_4OH to facilitate dissolution and to prevent effervescence. Shake until most of the DTPA is dissolved. Then add 79.06 g of NH_4HCO_3 and stir gently water. This solution is unstable with regard to pH and therefore, it is preferable to use a fresh solution each day.

Procedure

☆ Weigh 10 g of soil into 125 ml conical flask.

☆ Add 20 ml of AB-DTPA solution.

☆ Shake on a reciprocal shaker for 15 minutes with flasks kept open. Filter the extracts through Whatman No. 42 filter paper or its equivalent.

☆ Add concentrated HNO_3 @ 0.25 ml per 2 ml of the filtrate and shake for about 10 minutes to get rid of carbonate-bicarbonate matrix to prevent clogging of the capillary tip in the nebulizer.

☆ This solution is ready for simultaneous multi-element determination on ICAP.

Interpretation

The available amounts of different nutrients extracted by AB-DTPA are interpreted on the basis of following index values:

Table 5.1: Critical Limits (ppm) for AB-DTPA Extractable Nutrients

Status	P	K	Zn	Cu	Fe	Mn
Low	<3	<60	<0.9	<0.5	<2.0	<1.8
Medium	4.4	61-120	1.0-1.5	-	2.1-4.0	-
Adequate	8-11	>120	>1.5	>0.5	>4.0	>1.8

5.2 Nitrogen Analyzer as a Tool for Nitrogen Estimation

Principle

During digestion the nitrogen in the sample is converted to ammonium sulphate

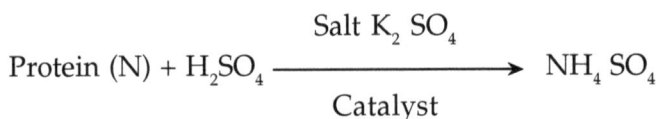

$$\text{Protein (N)} + H_2SO_4 \xrightarrow[\text{Catalyst}]{\text{Salt } K_2SO_4} NH_4SO_4$$

Distillation

The principle is to convert ammonium (NH_4^+) ions into ammonia (NH_3) by using alkali (NaOH) and thereafter steam distill it into receiver flask containing boric acid and titrate it against standard acid.

Titration

The boric acid is back titrated with a standard acid, preferably N/50 H_2SO_4 or HCl in routine soil analysis. At the end point the blue colour just disappears. One drop in excess turns the solution in original colour of boric acid.

Regents

Sodium Hybroxide (40 per cent)

Boric Acid (4 per cent) with bromocresol green/methyl red indicator solution. The 4 solution is prepared by dissolving 400 g of boric acid in about 5-6 litre very hot deionized water and make the volume to about 9 liter mix and add more hot deionized water to volume of about 9 litre. Cool the solution and add 100 ml bromoceresol green indicator (100 mg in 100 ml of methanol) and 70 ml of methyl red solution (100 mg in 100 ml of methanol). Dilute to 10 litre with deionized water.

Hydrochloric Acid (0.1 N)

Concentrated Sulphuric Acid

Procedure

☆ Weigh 1 g of sample to an accuracy of 0.1 mg into a digestion tube.

☆ Add two Kjeltaks Cu 3.5 (7 g K_2SO_4 + 0.8 g $CuSO_4$ $5H_2O$).

☆ Carefully add 12 ml of concentrated H_2SO_4 and gently shake.

☆ Attach the exhaust system to the digestion tubes in the rack and set the water aspirator.

☆ Load the rack with exhaust into a pretreated block (420°C)

☆ After 5 minutes run down the water aspirator until the acid fumes are just contained within the exhaust head.

☆ Continue to digest until samples are clear with blue/green solution. This will normally be after 30-60 minutes depending on the sample type.

☆ Remove the rack of tubes with exhaust still in place and put in the sand to cool for 10-20 minutes.

☆ Carefully add 75 ml deionised water to the tubes.

☆ Add 25 ml of boric acid (4 per cent) to a conical flask and place it into the distillation unit and raise the platform so that the distillate outlet is submerged in the receiver solution.

☆ Place the digestion tube in distillation unit and close safety door.

☆ Dispense 50 ml of 40 per cent NaOH into the tube.

☆ Open the steam valve and distil for approximately 4 minutes

☆ After about 90 per cent of distillation time lower the distillate platform.

☆ At the end of the distillation cycle close the steam value.

☆ The receiver solution in the distillate flask will not be green indicating the presence of an alkali ammonia.

☆ Titrate the distillate with standardized HCl (0.1 N) until blue/end point achieved. Note the value of acid consumed in titration. A blank should also be run to compensate any contribution from the reagent used.

$$\% \, N = \frac{(S\text{-}B) \times N \times 14 \times 100}{\text{Weigh of sample}}$$

S: Titration volume of sample

B: Titration volume of blank

N: Normality of acid used

Chapter 6

Analysis of Irrigation Water

6.1 Analysis of Irrigation Water

Every naturally occurring water contains varying quantities of dissolved solids and gases and in a few cases, some amount suspended organic and/or inorganic effects. Hence the test to be carried out to assess the quality of water, depends on the purpose for which it is likely to be used and also on the nature of ions present and their proportion in which they occur with respect to each other. In the case of irrigation waters, the nature and concentration of various ions, particularly the proportion of divalent ions to the movovalent cations, are of great importance in assessing the quality as these ions affect the crop growth differentaly. The cations that are present are calcium, magnesium, potassium and sodium and the anions commonly found are chloride, sulphate, carbonate and bicarbonate. In addition to the above, boron and nitrate-nitrogen are also found in a few cases are estimated if suspected to the above cations and anions and based on their contents, their quality is fixed.

6.1.1 Collection of Water Samples

Water samples should be collected in either a glass or polythene stoppered bottle preferably in a transparent one. It should be thoroughly cleaned before use and should be rinsed 3 to 4 times with the water which is to be sampled. Porous corks and rubber corks should be avoided. Normally one liter of water sample would be sufficient for all analysis

6.1.1.1 Sampling of Water

If the water sample has to be collected from tubewell, hand pump etc., they must run for at least 15 minutes so that the sediments, precipitates already formed either on the surface of the well or in the pipelines due to drying of materials, are washed away and are prevented from contaminating the sample. If the water has to be collected from an open well, a bottle of bucked with the lid closed, may be lowered to at least 2 to 3 meters below the surface without opening the lid. The water sample should be collected preferably from the centre of the well and never near the walls of the well.

If the source of irrigation water is an open tank, canal or river, the sample should be drawn from a spot in the midstream, away from the sides with a clean bucked previously rinsed with the same water. Then a proper label should be prepared indicating the source, name of the owner, crops to be raised, nature of soil and any other information like depth, problems encountered if any, in earlier seasons (or uses). The water sample should be analysed immediately and in case a few days of delays in taking up the analysis is inevitable, then 2 or 3 drops to tolune may be added to prevent bacterial activity.

6.2 Analysis of Waters

The water samples are analysed for the following parameters pH, total soluble salts (electrical conductivity), calcium, magnesium, sodium, potassium, carbonate, bicarbonate, chloride, sulphate, boron and nitrate-nitrogen.

6.2.1 pH

A water sample (approximately 100 ml) may be taken in a clean beaker and the pH is determined with the help of a pH meter. If large number of samples are being estimated then the buffers should be introduced after each batch of 10 or 15 samples and checked. If any deviations are observed then the instrument should be calibrated again and then only further analysis taken up.

6.2.2 Total Soluble Solids

6.2.2.1 Gravimetric Method

Take known amount of the water sample in a previously weighed potash basin and evaporate it to dryness on a water bath. In this analysis, use of silica basins or heavy weight containers should be avoided as it will be difficult to observe small differences in weights, caused by the solids in water. The potash basin with the salt of the water (dried water sample, after evaporation) should be kept in an air oven for one hour, then transferred to a dessicator and cooled. After it has attained the room temperature, find out the weight the basin plus the dried water sample. From these weights, calculate the total solids present in the water sample.

Calculation

 ☆ Aliquot of water sample = V ml

 ☆ Weight of empty potash basin = W_1 g

 ☆ Weight of potash basin + water = W

 ☆ Sample (after evaporation) = W_2 g

 ☆ Weight of total solids = (W_2-W_1) g

 ☆ Total solids (ppm) = $\dfrac{(W_2-W_1) \times 10^6}{V}$

6.2.3 Electrical Conductivity

The amount of total solids may assessed by the electrical conductivity of the water sample using conductivity meter. Clean the cell of the conductivity meter and fill the same with water sample after rinsing the cell atleast twice with the test sample. Find out the electrical conductivity and express in dsm^{-1}. Introduce check sample of known conductivity before and after every 10 or 15 samples and check the accuracy of the conductivity meter. The electrical conductivity is related to the total soluble salt content of irrigation waters.

6.2.4 Carbonates and Bicarbonates

Principle

The carbonate and bicarbonate ions are present as salts of strong bases and the alkaline in nature. These ions are estimated by titration with a standard acid using phenolphthalein and methyl orange as indicators. In the first part of the titration using phenolphthalein as indicator, the carbonates are converted to bicarbonate. In the second part of the titration using methyl orange as indicator, bicarbonates that have been formed is neutralised by the acid.

$$2Na_2CO_3 + H_2SO_4 \longrightarrow NaHCO_3 + Na_2SO_4$$

$$2NaHCO_3 + H_2SO_4 \longrightarrow Na_2SO_4 + 2H_2O + 2CO_2$$

Reagents

Sulphuric Acid (0.1 N)

Phenolphthalein

Methyl Orange

Procedure

☆ Pipette out 50 ml of the water sample into a 250 ml conical flask.

☆ Add 2 drops of phenolphthalein indicator

☆ Titrate against 0.1 N sulphuric acid till the disappearance of the pink colour

☆ Record the volume of standard acid consumed.

☆ To the colourless solution add 1 to 2 drops of methyl orange indicator

☆ Titrate till the presence of yellow colour.

☆ Note the volume of acid consumed

Observations and Calculation

☆ Volume of water taken = V ml

☆ Volume of 0.1 N sulphuric acid used with phenolphthalein = X ml

☆ Volume of 0.1 N sulphuric acid used with methyl orange = Y ml

☆ Volume of 0.1 N sulphuric acid required to neutralise bicarbonate alone = (Y-X) ml

☆ Volume of 0.1 sulphuric acid required to neutralise the carbonate alone = (2 x X) ml

(i) Carbonate

☆ 1 ml of 0.1 N sulphuric acid = 0.003 g carbonate

☆ (2 x X) ml of 0.1 N sulphuric acid = 0.003 x (2X) g carbonate

☆ Present in V ml of sample

Hence in 10^6 ml of water, carbonate (ppm)

$$= \frac{0.003 \times 2X \times 10^6}{V}$$

$$\text{Carbonate (me/liter)} = \frac{0.003 \times 2X \times 100 \times 1000}{V \times 30}$$

$$= \frac{2 X \times 100}{V}$$

(ii) Bicarbonate

☆ 1 ml of 01 N sulphuric acid = 0.0061 g of HCO_3

☆ (YxX) ml of 0.1 N sulphuric acid = 0.0061 x (Y-X) g of HCO_3

Present in V ml of water

Hence in 10^6 ml of water, bicarbonate (ppm)

$$= \frac{0.0061 \times (Y-X) \times 10^6}{V}$$

Bicarbonate (me/liter)

$$= \frac{0.0061 \times (Y-X) \; 100 \times 1000}{V \times 61}$$

$$= \frac{(Y-X) \times 100}{V}$$

6.2.5 Chlorides

Principles

Chloride are invariably present in all natural waters and contribute to its salinity. The chloride content is estimated by Mohr's method wherein the chloride is precipitated as silver chloride using silver nitrate. The excess of silver nitrate combines with potassium chromate that has been added as indicator to form a reddish brown coloured precipitate.

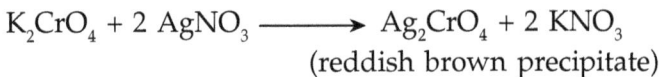

$$NaCl + AgNO_3 \longrightarrow AgCl + NaNO_3$$

$$K_2CrO_4 + 2\,AgNO_3 \longrightarrow Ag_2CrO_4 + 2\,KNO_3$$
$$\text{(reddish brown precipitate)}$$

Reagents

Silver Nitrate (0.02 N)

Potassium Chromate

Procedure

☆ Pipette out 50 ml of water sample into a clean white porcelain basin.

☆ Titrate against 0.02 N silver nitrate solution after adding potassium chromate

☆ Record the volume of standard silver nitrate used.

Calculation

☆ Volume of water taken = V ml

☆ Volume of 0.2 N silver nitrate used = A ml

☆ 1 ml of 0.02 N silver nitrate = 0.0071 g of Cl^{-1}

☆ A ml of 0.02 N silver nitrate = 0.00071 x Ag of Cl^{-1}

Present in V ml of water

Hence in 10^6 ml of water the chloride content (ppm)

$$= \frac{0.00071 \times A \times 10^6}{V}$$

$$Cl^{-1} \text{ (me/liter)} = \frac{0.00071 \times A \times 1000}{V} \times \frac{1000}{35.5}$$

6.2.6 Sulphate

Gravimetric Method

Principle

Sulphate present in the water sample either as sodium sulphate or calcium sulphate or magnesium sulphate. Sulphate in the water can be estimated gravimetrically or volumetrically. In the gravimetric method, the sulphate is precipitated as barium sulphate by adding barium chloride to the water sample in acidic solution. The precipitated barium sulphate is filtered, washed, ignited and weighed.

Reagents

Concentrated Hydrochloric Acid

Barium Chloride (10 per cent)

Methyl Orange

Procedure

☆ Take 100 ml of the water sample in a conical flask.

☆ Add a few drops of methyl orange and 1 ml of concentrated hydrochloric acid

☆ Heat to boiling and add an excess of barium chloride drop by drop with constant stirring.

☆ Allow it to stand on a water bath until the volume is reduced to 50 ml

☆ After cooling, the precipitate of barium sulphate is filtered through Whatman filter paper No. 3 (ashless) and washed with distilled water until it is free of chloride.

☆ Then fold the filter paper, keep it in a weighed silica crucible, ignite slowly initially then at high temperature.

☆ Then cool the silica crucible with the precipitate and record the weight of barium sulphate precipitate.

Observations and Calculation

☆ Volume of water = V ml

☆ Weight of empty silica crucible = W g

☆ Weight of slicia crucible + precipitate = W_2 g

☆ Weight of barium sulphate precipitate = $(W_2 - W_1)$ = W g

☆ Present in V ml of water

Hence amount of barium sulphate in

$$1000 \text{ ml} = \frac{W \times 1000 \text{ g}}{V}$$

$$\text{Weight of sulphate in 1000 ml} = \frac{\text{W} \times 1000 \times 0.412 \text{ g}}{\text{V}}$$

$$\text{Amount of sulphate in } 10^6 \text{ ml} = \frac{\text{W} \times 10^6 \times 0.412 \text{ g}}{\text{V}}$$

6.2.7 Boron

Principle

Normally waters intended are not tested for boron or nitrate nitrogen unless they are suspected to be present in appreciable quantities. Water particularly those from entrapped underground sources may contain excessive amount of boron. Organic compounds that form coloured complexes with boron as borate, Curcumin, a dye extracted from turmeric is preferred as it gives a sharp colour differentiation. For boron estimation an acid medium is preferred. It nitrate are present more than 90 ppm, they can be eliminated by evaporating the water to dryness after adding 10 ml of saturated solution of pure calcium hydroxide followed by ignition and extraction with 0.05 N hydrochloric acid. Only boron free glasswares like ordinary soda-glass or slicia or polythene wares should be used for boron analysis.

Reagents

Hydrochloric Acid (0.05 N)

Ethyl Alcohol (85 per cent boron free)

Curcumin Oxalic Acid Solution

Standard Boron Solution (100 ppm)

Apparatus

Spectrophotometer

Procedure

☆ Take a known quantity of water sample in 250 ml beaker

☆ Add 0.05 N hydrochloric acid till it is acidic.

☆ Add 4 ml of curcumin–oxalic acid reagent and evaporate to dryness at 55-60°C on a hot plate or water bath.

☆ Keep it for another 15 minutes at that temperature.

☆ After cooling, mix the residue thoroughly with 25 ml of 95 per cent ethyl alcohol to extract the rose cyamine coloured complex and filter it through Whatman No. 1 filter paper.

☆ Measure the intensity of the colour at 540 nm by spectrophotometer.

Calculation

$$\text{Boron (ppm)} = \frac{\text{μg of boron read from standard curve}}{\text{ml of water sample}}$$

6.2.8 Nitrate Nitrogen

Principle

Normally waters do not contain appreciable amounts of nitrate-nitrogen. Beneficial effects of nitrate–nitrogen have been reported in brackish waters. It is estimated by the phenol disulphonic acid method. A nitrate in dry condition, in the presence of concentrated sulphuric acid, causes nitration of phenol disulphonic acid, giving intense yellow colour compex in alkaline medium which can be measured colorimetrially at 420 nm.

Reagents

Calcium Sulphate

Ammonia (1: 1)

Phenol Disulphonic Acid

Procedure

☆ Take 50 ml of water sample in a porcelain dish or beaker

☆ Evaporate to dryness on a water bath and add 3 ml of phenol disulphonic acid and allow it to react for 10 minutes.

☆ Add 15 ml of distilled water and stir well with a glass rod.

☆ Transfer the contents to a 100 ml volumetric flask and add 1: 1 ammonia slowly with mixing till the solution is alkaline in reaction.

☆ Add 2 ml of ammonia and make up the volume to 100 ml with distilled water.

☆ Read the colour intensity at 420 nm by spectrophotometer.

Calculation

$$\text{Nitrate–nitrogen (ppm)} = \frac{\mu g \text{ of } NO_3\text{-N in test solution}}{\text{Volume of water sample}}$$

6.2.9 Calcium and Magnesium

Principle

Calcium and magnesium treated with EDTA to form a

complexed. Calcium and magnesium are estimated by the erichrome black T indicator. Erichrome black T indicator gives a wine red colour complex with magnesium but changes into sky blue when all the magnesium when all the magnesium ions have been removed by the versenate. Calcium is estimated using murexide indicator which forms an orange red complex of calcium at pH 12 and gets converted into a red-violet colour when all the calcium ions have been completely complexed by the versenate.

Reagents

Ethelene Diammine Tetra Acetic Acid (0.01 N)

Ammonium Chloride–Ammonium Hydroxide Buffer (pH 10)

Erichrome Black T

Murexide (ammonium purpurate)

Sodium Hydroxide (4 N)

Standard Calcium Chloride (0.01 N)

Procedure

Calcium + Magnesium

☆ Take 25 ml of water sample in a porcelain dish and dilute to with 25 ml with distilled water.

☆ Add one ml of ammonium chloride–ammonium hydroxide buffer solution and add 3 to 4 drops of erichrome black T.

☆ Titrate against 0.01 N EDTA (versenate) solution, till the colour changes from wine red to blue or bluish green.

☆ At the end, no tinge of red colour should be seen. Note the volume of EDTA used.

6.2.9.1 Calcium

☆ Take the 25 ml in porcelain dish and dilute it with 25 ml of distilled water.

☆ Add 5 drops of 4 N sodium hydroxide and 50 mg of purpurate indicator.

☆ Titrate with standard EDTA solution till the orange red colour change to purple.

☆ Also run a blank in the similar manner

Calculation

☆ Volume of water samples = 25 ml

☆ Volume of 0.01 N EDTA used for (Ca + Mg) = A ml

☆ Volume of 0.01 N EDTA used for calcium alone = B ml

☆ Volume of 0.01 N EDTA used for magnesium alone = (A-B) ml

Calcium

☆ 1 ml of 0.01 N EDTA = 0.002 g of Ca

☆ B ml of 0.01 N EDTA = 0.02 x B g of Ca.

☆ Present in 25 ml of water, hence

In 1000 ml of water

$$= \frac{0.0002 \times B \times 1000 \text{ g}}{25} \text{ of Ca/litre}$$

In 10^6 ml of water $= \dfrac{0.0002 \times B \times 10^6 \text{ g}}{25}$ of Ca

$$Ca\ (me/l) = \frac{0.002 \times B \times 1000 \times 1000}{25 \times 20}$$

$$= \frac{B \times 10}{25}$$

6.2.9.2 Magnesium

☆ 1 ml of 0.01 N EDTA = 0.00024 g of Mg

☆ (A-B) ml of 0.01 N EDTA = 0.0024 x (A-B) g of Mg

☆ Present in 25 ml of water. Hence

1000 ml of water contains

$$= \frac{0.00024 \times (A-B) \times 1000\ g}{25}\ of\ Mg$$

10 ml of water contains

$$= \frac{0.00024 \times (A-B) \times 10^6\ g}{25}\ of\ Mg$$

$$Mg\ (me/l) = \frac{0.0024 \times (A-B) \times 1000 \times 1000}{25 \times 24}$$

$$= \frac{(A-B) \times 10}{25}$$

6.2.10 Sodium and Potassium

Principle

The amount of sodium and potassium in good quality water present is very little. But in saline and brackish waters, the concentration of sodium may be quite high and the electrical conductivity of the waters will also be more than 1 mmhos/cm. The sodium and potassium is estimated by flame photometer. Sodium emits a bright yellow colour radiation at 589 nm. Similarly, potassium also emits a radiation which can be read by flame photometer. The intensity of emission is proportional to the concentration of sodium and potassium.

Apparatus

Flame photometer

Procedure

☆ Introduce the water sample in the flame photometer.

☆ Read the flame photometer reading.

☆ Find out the concentration of sodium from the standard curve.

Calculation

☆ me. Na/liter = concentration of Na obtain from standard curve (ppm) /equivalent weight

= ppm/22.99

6.2.10.1 Potassium

Principle

Potassium gives a violet colour radiation at 404.4 nm. The amount of potassium is estimated at 404.4 nm with blue filter.

Apparatus

Flame photometer

Procedure

☆ Introduce the water sample in the flame photometer.

☆ Read the flame photometer reading.

☆ Find out the concentration of potassium from standard curve.

Calculation

☆ me. K/liter = concentration of K obtain from the standard curve (ppm)/equivalent weight

= ppm/39.

6.2.11 Residual Sodium Carbonate (RSC)

The residual sodium carbonate gives a measure of the sodium hazard. Residual Sodium Carbonate is calculated using the amount of carbonate, bicarbonate, calcium and magnesium present in the waters.

$$\text{RSC (me./litre)} = (CO_3^{-2} + HCO_3^-) - Ca^{2+} + Mg^{2+})$$

Rating (me/l)	:	Interpretation
<1.25	:	Safe
1.25-2.50	:	Marginal
>2.50	:	Unsafe

6.2.12 Biochemical Oxygen Demand (BOD)

Biochemical oxygen demand is defined as the amount of oxygen requied by microorganisms to stabilize biologically decomposable organic matter in a waste under aerobic condition. The Biochemical oxygen demand test is generally performed to determine.

1. The degree of pollution in lakes and streams at any time

2. The pollutional load of waste waters.

BOD is evaluated by measuring oxygen concentration in sample, idometrically before and after incubation in the dark at 20°C for 5 days.

Apparatus

BOD incubator

BOD bottles

Reagents

DOB-free Water: Pass the deionized glass distilled water through a column of activated carbon and redistill it.

Phosphate Buffer Solution: Dissove 42.5 g potassium dihydrogen phosphate in 700 ml BOD free water and add 8.8 g NaOH. Adjust the pH 7.2. Add 2 g ammonium sulphate and dilute to 1 litre with BOD free water.

Magnesium Sulphate Solution: Dissolve 82.5 g of magnesium sulphate in BOD free distilled water and make the volume to 1 liter.

Calcium Chloride Solution: Dissolve 27.5 g of anhydrous calcium chloride in BOD-free distilled water and make the volume to 1 liter.

Ferric Chloride Solution: Dissolve 0.25 g of ferric chloride in 1 liter of BOD free distilled water.

Sulphuric Acid (1 N): Add 2.8 ml of concentrated sulphuric acid to 100 ml of BOD free distilled water.

Sodium Hydroxide Solution (1N): Add 4 g of sodium hydroxide in BOD free distilled water and make the volume 100 ml.

Allythiourea Solution: Dissolve 500 g of allythiourea in distilled water and make the volume 1 liter with BOD free water.

Procedure

Preparation of Dilution Water

☆ To prepare synthetic dilution water, aerate the required volume of BOD free distilled water in a glass container by bubbling compressed air for 1 to 2 day to attain dissolved oxygen saturation. After saturation keep at 20°C for at least one day.

☆ Add 1 ml each of phosphate buffer solution, magnesium sulphate solution, calcium chloride solution and ferric chloride solution.

Dilution of Sample

☆ Adjust the pH of sample to 7.0 by 1 N sulphruic acid or 1 N sodium hydroxide solution.

☆ Fill two sets of BOD bottles of either 125,250 or 300 ml with diluted water.

☆ Add 1 ml of allythiourea solution to each bottle

☆ Determine the dissolved oxygen content (D_O) in one set immediately following the Winkler's method of oxygen estimation. Incubate the other set in BOD incubator after 5 days determine immediately their dissolved oxygen content (D_5).

Calculation

$$BOD_5 \ (mg/l) = (D_O\text{-}D_5) \times \text{Dilution water}$$

where, D_O: initial dissolved oxygen in the sample (mg/l); and D_5: Dissolved oxygen left out in the sample after 5 days incubation (mg/l).

6.2.13 Chemical Oxygen Demand (COD)

Chemical oxygen demand (COD) measure the amount of oxygen required for oxidation of organic compounds present in water by chemical reaction involving oxidizing substances such as potassium dichromate and potassium permanganate. The test is widely used to determine.

1. The degree of pollution in water and their self purification capacity.

2. Pollution load.

Most of the organic matter decomposed and produces carbon dioxide and water when boiled with a mixture of potassium dichromate and sulphuric acid. A sample is refluxed with a known amount of potassium dichromate in sulphuric acid and the excess of dichromate is titrated against ferrous ammonium sulphate (FAS). The amount of dichromate consumed is proportional to the oxygen required to oxidize the organic matter.

Apparatus

COD reflux unit

Hot water bath

Reagents

Potassium Dichromate Solution (0.25 N): Dissolve 12.259 of AR grade potassium dichromate, dried at 103°C in distilled water. Add 120 ml sulphuric acid to this and dilute to 1 litre.

Silver Sulphate (dry powder)

Murcuric Sulphate (dry powder)

Concentrate Sulphuric Acid

Ferroin Indicator: Dissolve 0.695 g of ferrous sulphate and 1.485 g of 1, 10-phenonthroline in 100 ml of distilled.

Standard Ferrous Ammonium Sulphate Solution (0.25 N): Dissolve 98 g of ferrous ammonium sulphate in distilled water, add 20 ml of sulphuric acid, cool and dilute to 1 liter by distilled water. Standardize the solution, dilute 25 ml of potassium dichromate to 250 ml with distilled water, add 50 ml of sulphruic acid, cool and add 5-6 drops of ferrion indicator and titrate with ferrous ammonium sulphate solution. At the end blue green colour changes to reddish green. The normality of FAS is calculated as:

$$\text{Normality of FAS} = \frac{\text{Volume of } K_2Cr_2O_7 \text{ (ml) x 0.25}}{\text{Volume of FAS (ml)}}$$

Procedure

☆ Take 20 ml of water sample in reflux unit.

☆ Add 10 ml of potassium dichromate solution, a pinch of silver sulphate and mercuric sulphate and 30 ml of sulphuric acid.

☆ Attach Liebig candancer to the mouth of flask and heat the flask for at least 2 hour to reflux the contents.

☆ Cool the flask and dilute to 150 ml with distilled water.

☆ Add 2-3 drops of ferroin indicator and titrate with ferrous ammonium sulphate solution. At the end blue green colour changes to reddish colour.

☆ Run a blank simultaneously in the similar manner.

Calculation

$$COD \ (mg/l) = \frac{(B\text{-}T) \times \times N \times 1000 \times 8}{Volume \ of \ sample \ (ml)}$$

where, T = volume of titrant (FAS) used against sample (ml): B= volume of titrant (FAS) used against blank (ml) N = normality of ferrous ammonium sulphate, equivalent weight of oxygen is 8.

Chapter 7
Laboratory Facilities

7.1 Laboratory Equipments

A modern soil testing laboratory for soil and plant analysis uses techniques which rely on electronic instruments such as pH meter, conductivity meter. Flame photometer, spectrophotometer and other advanced instruments like, Nitrogen analyzer and inductive coupled plasma atomic emission spectrophotometer (ICP-AES) shaking and centrifuging are used exclusively with the aid of electrical machines and devices (Figure 7.1).

7.2 Glassware and Plasticware

In soil testing laboratory the glassware which are most commonly used for chemical analysis are Borosilicate glass because of its low coefficient of expansion when heated except for the determination of boron (B), where boron free glassware are used for the purpose. The most commonly used glassware plastic were are given in (Figure 7.2) these mainly includes

Figure 7.1: Flame Photometer

Figure 7.2: Atomic Absorption Spectrophotometer

Figure 7.3: Muffle Furnace

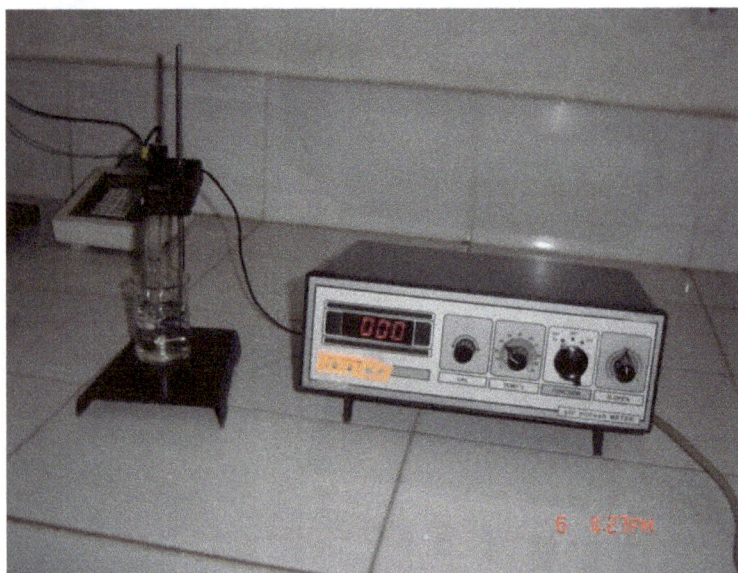

Figure 7.4: pH Meter

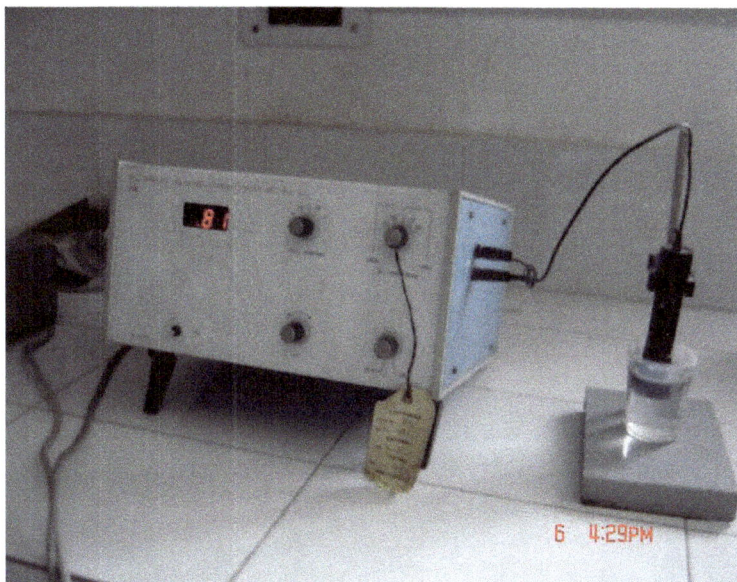

Figure 7.5 : Conductivity Meter

Figure 7.6: UV Visible Spectrophotometer

Figure 7.7: Mechanical Shaker

Figure 7.8: Inductively Coupled Plasma Atomic Emission Spectrophotometer (ICAP-AES)

Figure 7.9: Nitrogen Analyser

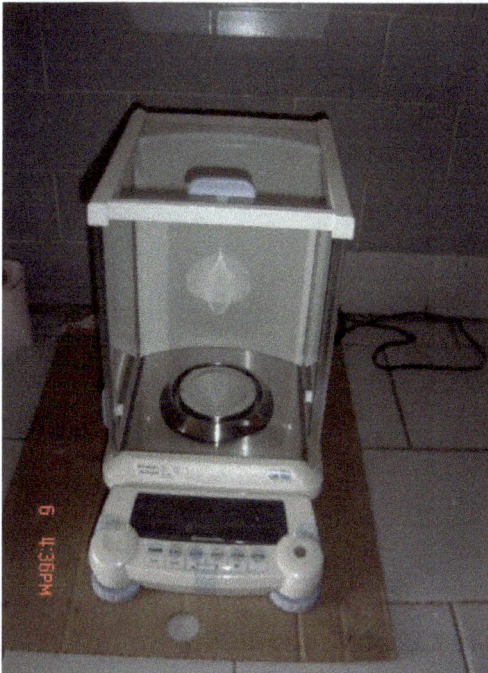

**Figure 7.10:
Analytical Balance**

Figure 7.11: Centrifuge

Figure 7.12: Conical Flask

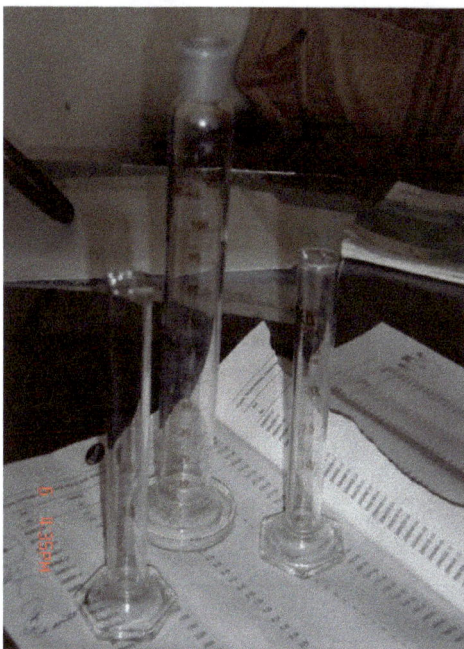

**Figure 7.13:
Measuring Cylinder**

**Figure 7.14:
Volumetric Flask**

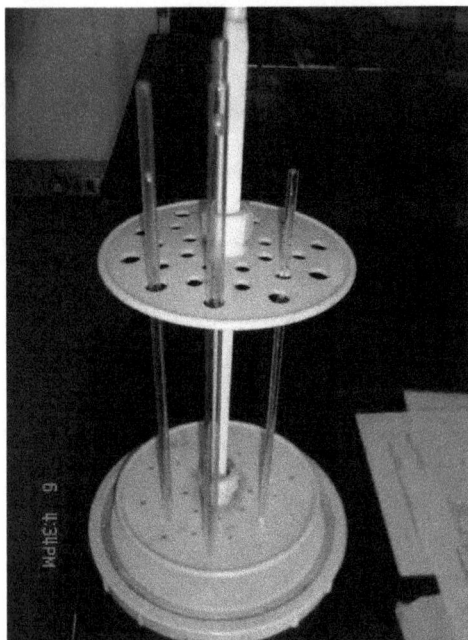

Figure 7.15: Pipetts

7.3 Chemical and Solutions

Generally, a high grade chemical, guaranteed by manufacturer to conform to stated quality, is advisable to use in all the analytical works. The chemicals which absorb water from the air or lose water, if possible their use should be avoided or anhydrous salts be used which are more stable than one containing water of crystallization.

The care and accuracy should be taken for preparation of reagents/solution. Measurement of volumes of liquid is always be made by using graduated volumetric glasswares.

The standard solution must be prepapred with greater care from the purest chemicals and high grade distilled or deionized water. Also accurate analytical balance of high precision) and accurately graduated volumetric glasswares should always be used for dependable results.

APPENDICES

Appendix–I

Nutrient Content of Some Important Fertilizers and Manures (%)

Sl.No.	Fertilizer/Manures	N	P_2O_5	K_2O
1.	Ammonium Sulphate	20.6	–	–
2.	Ammonium Sulphate Nitrate	26.0	–	–
3.	Ammonium Nitrate	33.0	–	–
4.	Ammonium Chloride	26.0	–	–
5.	Calcium Ammonium Nitrate	20.6	–	–
6.	Sodium Nitrate	16.0	–	–
7.	Urea	46.0	–	–
8.	Calcium Cyanamide	15.5	–	–
9.	Anhydrous Ammonia	82.0	–	–
10.	Mono Ammonium Phosphate	11.0	48.0	–
11.	Dia ammonium Phosphate	18.0	46.0	–
12.	Ammonium Phosphate Sulphate	16.0	20.0	–
13.	Single Superphosphate	–	16.0	–
14	Triple Superphosphate	–	48.0	–
15.	Dicalcium Phosphate	–	32.0	–
16.	Basic slag	–	20.25	–
17.	Bone meal Raw (steamed)	–	23.30	–
18.	Calcium Metaphosphate	–	62.63	–
19.	Potassium Chloride	–	–	60

Contd...

Contd...

Sl.No.	Fertilizer/Manures	N	P_2O_5	K_2O
20.	Potassium Sulphate	–	–	48
21.	Potassium Nitrate	13	–	44
	Manures			
22.	Farm yard manure	1.2	0.6	1.2
23.	Ship manure	1.9	1.3	2.3
24	Poultry manure	0.9	1.8	0.6
25.	Pig manure	3.7	3.3	0.4
26.	Compost (municipal)	0.4	0.6	0.4
27.	Compost leaf	1.0	0.7	1.0

Appendix–II

Important Volumetric Primary Standards

Primary Standard	Formula	Eq.wt/ Mol. Wt.	Eq. wt	Mol. Wt
S–Diphenylguandine	$NH: C (NC_6H5)_2$	1	211.26	211.26
Meruric oxide	HgO	½	108.31	216.62
Potassium acid carbonate	$KHCO_3$	1	100.12	100.12
Potassium iodate	KIO_3	1/6	36.67	204.02
Sodium carbonate	Na_2CO_3	½	53.00	106.00
Sodium oxalate	$Na_2C_2O_4$	½	67.00	134.00
Sodium tetraborate (Borax)	$Na_2B_4O_7, 10H_2O$	½	190.72	381.44
Benzoic acid	$C_6H_5 COOH$	1	122.13	122.13
Hydrazine sulphate	$N_2H_4H_2SO_4$	½	65.07	130.14
Oxalic acid crystals	$H_2C_2O_4 . 2 H_2O$	½	63.03	126.06
Potassium acid oxalate	KHC_2O_4	1	128.13	128.13
Potassium acid phthalate	$KHC_8 H_4O_4$	1	204.23	204.23
Potassium acid tartarate	$KHC_4 H_4O_6$	1	188.18	188.18
Potassium tetraoxalate	$KH_3(C_2O_4)_2 . 2H_2O$	1/3	84.73	254.20
Ferrous sulphate	$FeSO_4.7 H_2O$	1/3	278.03	278.03
Ferrous ammonium sulphate	$FeSO_4. (NH_4) 2SO_4. 6H_2O$	1	392.16	392.16

Contd...

Contd...

Primary Standard	Formula	Eq.wt/ Mol. Wt.	Eq. wt	Mol. Wt
Iron wire	Fe	1	55.85	55.85
Potassium ferrocyanide	$K_4Fe(CN)_6 \cdot 3H_2O$	1	422.42	422.41
Silver	Ag	1	107.88	107.88
Arsenious oxide	As_2O_3	¼	49.455	197.82
Copper	Cu	1	53.54	53.54
Hydrazine sulphate	$N_2H_4 \cdot H_2SO_4$	¼	34.531	130.124
Iodine (resublimed)	1	1	126.91	126.91
Iodine cyanide	ICN	½	76.465	152.93
Potassium bormate	$KbrO_3$	1/6	27.836	167.016
Potassium dichromate	$K_2Cr_2O_7$	1/6	49.037	294.222
Potassium ferricyanide	$K_3 Fe (CN)_6$	1	329.26	329.26
Sodium thiosulphate	$Na_2S_2O_3 \cdot 5 H_2O$	1	248.2	248.21
Potassium bromide	KBr	1	119.02	119.02
Potassium chloride	KCl	1	74.56	74.56
Sodium chloride	NaCl	1	58.45	58.45
Mercury	Hg	½	100.31	200.62
Silver nitrate	$AgNO_3$	1	169.89	169.89

Appendix-III

Important Gravimetric Conversion Factors

Multiply the Weight of	By this Factor	To Obtain Weight of
Al	1.8895	Al_2O_3
Al_2O_3	0.52913	Al
Ba	1.6994	$BaSO_4$
$BaSO_4$	0.58845	Ba
$BaSO_4$	0.13737	S
$BaSO_4$	0.41154	SO_4
Ca	2.4973	CaO_4
Ca	4.2959	$CaSO_4.2\ H_2O$
$CaCO_3$	0.4004	Ca
$CaSO_4.2\ H_2O$	0.23277	Ca
CaO	0.715	Ca
CaC_2O_4 (oxalate)	0.3128	Ca
$CaSO_4 . 2\ H_2O$	0.5579	SO_4
$CaCO_3$	0.43971	CO_2
AgCl	0.24737	Cl
Cl	4.0426	AgCl
KCl	0.47557	Cl
Cl	2.1027	KCl
NaCl	0.60664	Cl
Cl	1.6484	NaCl

Contd...

Contd...

Multiply the Weight of	By this Factor	To Obtain Weight of
Cu	3.9296	$CuSO_4.2H_2O$
$CuSO_4.5H_2O$	0.25448	Cu
$FeSO_4 (NH_4)_2SO_4.5H_2O$	0.14242	Fe
Fe	1.4297	Fe_2O_3
Fe_2O_3	0.69944	Fe
$FeSO_4. 7 H_2O$	0.20088	Fe
$FeSO_4. 7 H_2O$	0.28720	Fe_2O_3
MgO	0.60317	Mg
$Mg_2 P_2O_7$	0.2185	Mg
$MgSO_4. 7 H_2O$	0.09866	Mg
MgP_2O_7	0.36228	MgO
N	1.2159	NH_3
N	4.7168	$(NH_4)_2 SO_4$
N	4.4266	NO_3
NH_3	0.82245	N
NH_3	3.6407	NO_3
$NaNO_3$	0.1648	N
$(NH_4)_2 SO_4$	0.21201	N
NO_3	0.22591	N
NO_3	0.27467	NH_3
NO_2	0.30447	N
P	2.2914	P_2O_5
P	3.0661	PO_4
PO_4	0.32613	P
PO_4	0.7473	P_2O_5
P_2O_5	0.43642	P
P_2O_5	2.18	$Ca (PO_4)_2$
P_2O_5	1.3381	PO_4
$Ca (PO_4)_2$	0.45762	P_2O_5

Contd...

Contd...

Multiply the Weight of	By this Factor	To Obtain Weight of
$Ca(PO_4)_2$	0.1997	P
$Ca(H_2PO_4)_2^- H_2O$	0.2457	P
KH_2PO_4	0.2276	P
K	1.2046	K_2O
KCl	0.52443	K
K_2O	0.83015	K
K_2SO_4	0.44874	K
$KClO_4$	0.28219	K
AgCl	0.2473	Cl
$AgNO_3$	0.84371	AgCl
$AgNO_3$	0.2087	Cl
Na	1.5422	Cl
Na_2CO_3	1.2855	$CaSO_4$
NaCl	0.6066	Cl
NaCl	0.39336	Na
$NaHCO_3$	0.273767	Na
Na_2SO_4	0.32371	Na
S	5.3695	$CaSO_4 . 2H_2O$
S	7.2795	$BaSO_4$
S	2.995	SO_4
$BaSO_4$	0.13737	S
$BaSO_4$	0.41155	S
$CaSO_4. 2H_2O$	0.1862	SO_4
$CaSO_4 . 2H_2O$	0.5566	$CaSO_4 . 2H_2O$
SO_4	1.7812	S
SO_4	0.3217	Zn
ZnO	0.8033	ZnO
Zn	1.2447	Zn
$ZnSO_4 . 7H_2O$	0.22736	Zn
Zn	4.3982	$ZnSO_4. 7H_2O$

Appendix–IV

Suggested Rates and Sources of Secondary and Micronutrients for Foliar Application

Element	Lbs. Element per Acre	Suggested Sources
Calcium (Ca)	1–2	Calcium chloride or calcium nitrate
Magnesium (Mg)	1–2	Magnesium sulfate
Manganese (Mn)	1–2	Soluble manganese sulfate or finely ground manganese oxide
Copper (Cu)	0.5–1.0	Basic copper sulfare or copper oxide
Zinc (zn)	0.3–1.0	Zinc sulfare
Boron (B)	0.1–0.5	Soluble borate
Molybdenum (Mo)	0.0	Sodium molybdate
Iron (Fe)	1–2	Ferrous sulphate

*: Use a minimum of 30 gallons of water per acre.

Appendix–V

Standards of Quality of Water for Various Industries
(Limiting tolerance of impurities in ppm)

Impurity	Washing	Irrigation	Textiles	Sugar Factory	Pottery	Steam Boiler
Ca^{2+}	10	40	10	20	–	5
Mg^{2+}	5	20	5	10	–	5
Cl^-	500	100	100	20	20	100
SO_4^{2-}	500	200	100	20	20	100
Fe^{2+}	0.1	–	0.1	0.1	–	–
HCO_3^-	–	–	200	100	–	100

Appendix–VI

Concentrations of Acids and Ammonia Hydroxide

Reagent	Specific Gravity	Approxi- mately Normality	Per cent Purity	ml Required to Prepare 1 Liter of 1 N Solution
Acetic acid glacial	1.05	17.0	99	57.5
Hydrochloric acid	1.18	12.4	38	82.6
Hydro fluoric acid	1.15	28.8	48	36.0
Nitric acid	1.42	16.0	72	47.4
Perchloric acid	1.67	11.6	70	86.2
Phosphoric acid	1.75	48.0	85	22.7
Sulphuric acid	1.84	37.0	98	28.1
Ammonium hydroxide	0.91	13.0	28.33	67.6

References

American Public Health Association (1985). *Standard methods for the examination of waste and waste water.* 16th ed. Washington, D.C.

Arora, C. L. and Hundal, H,S. 1995. Evaluation of ammonium bicarbonate–DTPA as a multielement extractant for soils. National seminar on developments in soil Sci. 60th Annual convention of ISSS. P.A.U. Ludhiana 2-5 Nov. 1975.

Black, C.A. 1965. Methods of soil analysis, part I and II. Am. Soc. Agronomy Inc. Publishers, Madison, Wisconsin, USA.

Bouyoucons, G..J. 1927. The hydrometer as a new method for the mechanical analysis of Soils. *Soil Sci.* 23: 343-353.

Bray, R.H. and Kurtz, L.T. 1945. Determination of total organic and available forms of phosphate in soils. *Soil Sci.* 59: 39-45.

Datta, N.P., Khera, M.S. and Saini, T.R. 1962. A rapid colorimetric procedure for the determination of the organic carbon in soils. *J. Indian Soc.* 10: 67-74.

Ensminger, L.E. 1954. Some factors affecting the adsorption of sulphate by Alabama soils. *Proc. Soil Sci. Soc. Proc.* 19: 279-282.

Greig, J.L. 1953. Determination of available molybdenum of soils. *N.Z.J. Sci. Tech. Sect.* A-34: 405-414.

Gupta, U.C. 1979. Some factors affecting the determination of hot water soluble boron from Podzol sols using azomethine-H. *Can. J. Soil Sci.* 59: 247-249.

International Society of Soil Science 1929. Minutes of the first commission meetings, international congress of soil science, Washington 1927. *Proc. Inter Soc. Soil Sci.* 4: 215-220.

Jackson, M.L. 1973. *Soil chemical analysis.* Prentice Hall of India Pvt. Ltd. New Delhi.

Kenny, D.R. and Bremner, J.M. 1962. Chemical index of soil nitrogen availability. *Nature,* London, 211: 892-893.

Krishnamurti, G.S.R., Mahavir, A.V. and Sharma, V.A.K. 1970. Spectrophotometric determination of Fe with orthophenathroline. *Micro-chemical Journal* 15: 585-589.

Kundsen, D., Paterson, G.A. and pratt, P.F. 1982. Lithium, sodium and potassium pages 225-246. in A.L. Page *et al.,* eds. Methods of soil analysis. Part 2 2nd ed. Agronomy No. 9 American society of Agronomy, Madison, W.I.

Lindsay, W.L. and Norvell, W.A. 1978. Development of DTPA soil test zinc, manganese and copper. *Soil Sci. Soc. Am. J.* 42: 421-428.

Massoumi, A. and Cornfield, A.H. 1963. A rapid method for determining sulphate in water extracts of soils. *Analyst London* 8: 321-322.

Mehlich, A. 1978. Determination of Phosphorus, Potassium, Calcium, Mangesium, Sodium, Manganese and Zinc. Communication in Soil Science and Plant Analysis. 9: 477-492.

Olsen, S.R., Cole, C.V, Watanabe, F.S., Dean, L.A. 1954. Estimation of available phosphorus in soils by extraction with sodium bicarbonate US Department Agriculture Circular.

Palaskar, M.S., Babrekar, P.G and Ghose, A.B. 1981. A rapid analytical technique to estimate sulphur in soil and plant extract. *J. Indian Soc. Sci.* 29: 249-256.

Prasad, R. 1965. Determination of potentially availability nitrogen in soils a rapid procedure. *Plant and Soil* 23(2): 261-264.

Purvis, E.R. and Peterson, N.K. 1956. Methods of soil and plant analysis for molybdenum. *Soil Sci.* 81: 223-228.

Scholler Berger, C.J. and Simon, R.H. 1945. Determination of exchange capacity and exchangeable bases in soil–ammonium acetate method. *Soil Sci.* 59: 13-24.

Shaw, E. and Dean, L.A. 1952. Use of dithizone as an extractant to estimate the zinc nutrient status of soils. *Soil Sci.* 73: 341-347.

Shoemaker, H.E., McLean, E.O. and Pratt, P.F. 1961. Buffer methods for determining lime requirement of soil with appreciable amounts of extractable aluminium. *Proc. Soil Sci. Soc. Am.* 25: 274-277.

Soltanpour, P.N. and Schwab, A.P. 1977. A new soil test for simultaneous extraction of macro-and micronutrient in alkaline soils. *Commun. Soil Sci. Plant. Anal.* 8(3): 195-207.

Soltanpour, P.N. and Workman, S.M. 1979. Modification of the NH_4HCO_3 DTPA soil test to omit carbon black. *Commun Soil. Sci. Plant. Annal.* 10: 1411-1420.

Soltanpour, P.N., Workman, S.M. and Schwab, A.P. 1979. Use of inductively coupled plasma spectrometry for the simulateneous determination of macro- and micro-nutrients in NH_4CO_3–DTPA extracts of soils. *Soil Sci. Soc. Am. J.* 43: 75-78.

Subbiah, B.V. and Asija, G.L. 1956. A rapid procedure for the determination of available nitrogen in soils. *Curr. Sci.* 25: 259-260.

United States Salinity Laboratory Staff 1954. Diagnosis and improvement of saline and alkali soils. *U.S. Dept. Agr. Handb* 60 U.S. Govt. Printing Office, Washington, D.C. 160 p.

Walkley, A. and Black, I.A. 1934. An examination of the degtsareff method for determining soil organic matter and a proposed modification of the chromic acid tirtration methods. *Soil Sci.* 34: 29-38.

Watanabe, F.S. and Olsen, S.R. 1965. Test of ascorbic acid method for determining phosphorus in water and sodium bicarbonate extracts of soil. *Proc. Soil Sci. Soc. Am.* 29: 677-78.

Willand, H.H. and Greathous, L.H. 1917. The colorimeteric determination of manganese by oxidation with periodets. *J. Am. Chem. Soc.* 391: 2355-2366.

Wolf, B. 1971. The determination of boron in soil extracts, plant materials compost, manures, water and nutrient solutions. *Soil Sci. Pl. Anal.* 2: 363-374.

Index